CLEAN HANDS:
Clair Patterson's Crusade Against Environmental Lead Contamination

CLEAN HANDS:
Clair Patterson's Crusade Against Environmental Lead Contamination

Edited by
CLIFF I. DAVIDSON
Director, Environmental Institute
Carnegie Mellon University

Foreword by Nobel Laureate
Saul Bellow

Nova Science Publishers, Inc.
Commack, New York

Editorial Production: Susan Boriotti
Office Manager: Annette Hellinger
Graphics: Frank Grucci and John T'Lustachowski
Information Editor: Tatiana Shohov
Book Production: Donna Dennis, Patrick Davin, Christine Mathosian and Tammy Sauter
Circulation: Maryanne Schmidt
Marketing/Sales: Cathy DeGregory

Library of Congress Cataloging-in-Publication Data

Clean Hands: Clair Patterson's crusade against environmental lead contamination / Cliff I. Davidson, ed.; foreword by Nobel Laureate Saul Bellow.
p. cm.
Includes bibliographical references and index.
ISBN 1-56072-568-0
1. Patterson, Clair C. 2. Geochemists--United States--Biography. 3. Lead--Environmental aspects. I. Davidson, Cliff I.. II. Title.
QE514.62.P37C48 1998 98-21167
363.738'4--dc21 CIP

6080 Jericho Turnpike, Suite 207
Commack, New York 11725
Tele. 516-499-3103 Fax 516-499-3146
e-mail: Novascience@earthlink.net
e-mail: Novascil@aol.com
Web Site: http://www.nexusworld.com/nova

Printed in the United States of America

To Laurie Patterson

CONTENTS

FOREWORD

About fifteen years ago, I gave a course at Caltech on the early 20th century novel. The students who signed up for it were as mystified by Joseph Conrad as I was by jet propulsion. When we discussed *Typhoon* one of them said that he was bothered by the disorderly thinking of Captain MacWhirr. He began his analysis by assigning the Captain and his first mate to subset-a, the crew to subset-b, and the Chinese passengers to subset-c. Then he went to the blackboard and began to write out an equation.

"I'm afraid you've lost me," I said.

Then all the boy prodigies and I stared at one another in shocked silence.

At the center of the classroom there sat a white haired, close-cropped man of about my own age. I took him for a visitor, an auditor -- a member of the faculty, perhaps, with an hour to kill. He was lean and tall, rather Western looking. I remember wondering whether this man might possibly explain how these clever boys, elite engineers and technocrats in training, viewed the paleo-technic freighter Nan Shan, its old-fashioned literal-minded Captain, its engineer and all the Coolie passengers in the hold.

My visitor, whose attendance record was flatteringly perfect, was Clair Patterson, a geochemist, and a graduate of the University of Chicago. In the early thirties, the worst years of the Great Depression, both of us had been Chicago undergraduates, on opposite sides of the quadrangle -- the sciences and the humanities separated by lawns, trees and tennis courts.

Patterson introduced himself and invited me to lunch at the faculty club. We crossed the campus together. On all sides, dormitory windows were open and the air was hammered by competing rock bands. I suggested that this huge noise, cutting the students off from the rest of the world, was possibly an aid to the study of particle physics or molecular biology. Captain McWhirr was similarly sealed by the roar of the typhoon. He was able to maintain an inward stillness. But he wasn't doing intellectual work, was he. He concentrated on the responsibilities of command.

Patterson's answer was that these brainy students had been brought up on Mickey Mouse and Superman and that their picture of human life was inevitably a caricature.

During our lunches and our walks I came to understand quite a lot about Pat Patterson. William Jennings Bryan, speaking for the Free Silver Democrats exactly a century ago, cried out against the Cross of Gold.

Pat Patterson's cross was made of lead. Fascinated I listened to his description of the gradual contamination of the planet Earth by lead. I was dazzled by the originality of his researches and by their scope. He told me that for several millennia now the lead levels of sea water and of the air had risen steeply. Pattern had measured lead residues in human and animal bones, in arctic ice strata and was able to prove, I was persuaded, that the brains of civilized man had suffered lead damage as well. Moving to conquer nature as the Enlightenment had told us we must, man had fatally damaged the earth, had poisoned its air and its waters. Already our brains were affected by the lead wastes we had ingested, our descendants would be ever more seriously and progressively brain-damaged. Our conquest of nature, from Patterson's standpoint, was therefore not a triumph but a defeat.

This, as our friend saw it, was the true history of the earth, of life and of humankind.

I was always inclined to view Patterson's account as prophetic revelation. As such, I could appreciate it for its comprehensiveness and its somewhat poetic splendor. Perhaps I am already too seriously brain-damaged to review the scientific data underlying his conclusion.

Pat used to say consolingly that I had done remarkably well for a man of my lead-poisoned generation. In an earlier time I should have done far better.

I had occasionally thought that temperamentally I might have been better suited to the late middle ages, or to the 18th century.

Many great thinkers have held that we are sick creatures. Baudelaire wrote that modern man was like a hospital patient who believed that he would recover if only he were moved to another bed. Freud was convinced that the repression of our instinctual desires was the cause of our lifelong illnesses. Pat Patterson held that we were poisoned by the accumulation of lead deposited over centuries in earth, air and water.

He once told me how melancholy it made him to sit at the open doors of the Patterson barn in Iowa and look out over wheat fields ruined by many decades of chemical fertilizers and weed-killers.

When Patterson discovered that planetary lead levels were dangerously high, he did everything possible to publicize his findings. One of the notable, practical results of his efforts was the removal of lead from gasoline. A generous man of noble instincts, he was deterred neither by pressures from special interest groups nor by the fact that his career was often at risk.

Saul Bellow

PREFACE

In his book *Horace's Compromise*, Theodore Sizer calls attention to the failure of the American education system. He notes that "young people grow in idiosyncratic, variable ways, often unpredictably. Good teachers are essential to nourish this growth. No system or school site leadership, no treatment or intervention, no innovative program, approved textbook, or curriculum guide can overcome their influence, for good or ill." Sizer goes on to emphasize that teachers are vital to set standards, help students, and serve as models.

Professor Clair Patterson served many roles throughout his life, but perhaps none was more important than serving as a role model for the nation's young future scientists. His life's work is an outstanding example of how one person gave us new knowledge that had a positive impact on people throughout the world.

The primary purpose of this book is to enable our future scientists -- and current scientists -- to learn from this remarkable career. The themes of the career are simple enough to state: perseverance, dedication, a desire to achieve one's potential, and integrity above all else. What is not so simple is how these themes played out during Patterson's life. To explore this issue the story of Patterson is told from different perspectives by those who worked with him and knew him. It is a unique story, with plenty of adventure, detective work, and achievement in the face of seemingly insurmountable obstacles. It is also a story of success.

Before embarking on the story of Patterson's life, it is important to consider his personal philosophy of science. The key element of this philosophy was that scientific knowledge is completely objective: there is no "good" or "bad" knowledge. The role of the scientist is to discover this knowledge; the primary motivation is not social conscience but rather a personal desire to discover the truth. There may also be secondary reasons why an individual decides to embark on a scientific career, such as to provide new knowledge that can help people. However, a responsible scientist must understand the distinction between these two very different perspectives. In the first, scientific truth is a pursuit in its own right, a thing of beauty. In the second, scientific discoveries offer knowledge that can benefit society, and the work is driven by "utilitarian" concepts.

As was clear throughout Patterson's career, it is sometimes difficult to keep a boundary between objective scientific thought and utilitarian concepts. Nevertheless, such a boundary must be maintained: searching for the truth in the natural world becomes difficult if a particular set of human beliefs becomes the main motivation. It was Patterson's unswerving allegiance to science, not to a person or a political system, that enabled him to keep moving forward despite adversaries.

The message to young people today is clear: a career in science can be immensely rewarding, both in terms of the intrinsic rewards as well as the rewards of helping others. However, it is necessary to recognize that striving for the truth has its costs -- and the most important of these is perhaps the need to put the quest for the truth above personal feelings. Patterson's career is one of the best illustrations of this philosophy of science. With such a philosophy, it is indeed possible to change the world for the better.

Cliff Davidson

ACKNOWLEDGMENTS

The authors of this volume wish to acknowledge the generous assistance of several people. Robert Elias provided information on his associations with Patterson and valuable data on emissions of lead. Letters from the files of the journal *Archives of Environmental Health* cited in the Introduction were provided by Joel Tarr. Some ideas on the content of this book were suggested by Samuel Hays. Photographs of Clair Patterson for use throughout the book were provided by Laurie Patterson, Claude Boutron, Robert Duce, Dorothy Settle, and Lee Silver. Kathy Lima provided a copy of Clair Patterson's Bibliography and assisted with obtaining additional photographs, as well as arranging for reproduction of photographs and documents from the Memorial Service. Information for the table of leaded and unleaded fuel was obtained from Evelyn Mauss. A special thanks to Laurie Patterson for providing numerous letters and other documents with information that has been used throughout the book.

The editor appreciates the opportunity to present some of this material in classes at Carnegie Mellon, especially classes of Professors Indira Nair and James Garrett.

The current volume was motivated by *Earth and the Human Future: Essays in Honor of Harrison Brown* by Kirk Smith, Fereidun Fesharaki, and John Holdren. Copies of that book for all the authors of the current volume were kindly provided by Kirk Smith.

A special thanks is due to Megan Davidson for proofreading the manuscript.

About the Editor

Cliff I. Davidson is Professor of Civil and Environmental Engineering/Engineering and Public Policy and Director of the Environmental Institute at Carnegie Mellon University. He is active in the American Association for Aerosol Research (Vice-President 1998-1999 and President 1999-2000) and several other professional societies. Other books: J.O. Nriagu and C.I. Davidson, *Toxic Metals in the Atmosphere*, Wiley Publishing, 1986; C.I. Davidson and S.A. Ambrose, *The New Professor's Handbook*, Anker Publishing, 1994; E.S. Rubin and C.I. Davidson, *Introduction to Engineering and the Environment*, McGraw-Hill Publishing, under contract 1999.

INTRODUCTION

Clair Cameron Patterson (1922-1995) was one of the world's great scientists. He gave us the first accurate age of the earth, pioneered the concept of contamination control in handling environmental samples, and devoted his life to the study of lead pollution. How one individual could do so much in a lifetime -- and help the world in so many ways -- is a story that deserves to be told. There were two key factors that shaped this remarkable career.

The first factor was the state of the world when Patterson inherited it. A unique set of circumstances had developed that made the world ripe for change. Numerous pieces of evidence over many decades showed that lead was a toxic element and had to be controlled, but nothing was done until the situation had reached a crisis. The second factor was Patterson's personality. It took a special individual to recognize that there was a problem of lead pollution on a massive scale. Not only did Patterson discover the problem and recognize its importance, but in addition he fought conventional wisdom -- established by industries, government, and fellow scientists -- to achieve one of the greatest environmental successes of all time: his findings played a key role in influencing governments worldwide to remove lead from gasoline.

In this book, we tell the story of Clair Patterson from the perspectives of ten individuals who knew him. The chapters offer different perspectives, since each person had a different relationship with Patterson. Nevertheless, there are certain consistencies that emerge when all of these chapters are viewed together. The result is an important message: one individual *can* make a difference. Before delving into the chapters written by the various authors, let us consider the two major factors that helped to shape Patterson's career.

THE STATE OF THE WORLD WHEN PATTERSON INHERITED IT

Before Patterson's early work in the 1950's and 1960's, we knew little about how human activities pollute the environment. This was partly because we couldn't accurately measure pollutants in the air, water, food, and other media. In this section, we first consider the events that had resulted in worldwide lead contamination by the time Patterson began his work. We then examine the state of knowledge in the field of geochemistry that enabled Patterson to discover the ubiquitous nature of lead pollution. Finally, we summarize what was known about environmental lead in these early years to show the discrepancies that existed between Patterson's results and those of other scientists.

EVENTS THAT RESULTED IN WORLDWIDE LEAD CONTAMINATION

Lead has been used in many commercial products over the years: solder, paint, batteries, plumbing, and leaded fuel are just a few examples. The manufacture and use of these products inevitably give rise to emissions to the environment, and this has had an impact on public health. The product with the greatest emissions of lead to the environment in years past is leaded gasoline. In 1975, for example, this source accounted for over 60% of the global anthropogenic lead emissions to the atmosphere (Pacyna, 1986); the corresponding figure for the U.S. was 88% in this same year, according to the U.S. Environmental Protection Agency (1977). Therefore, we focus on leaded gasoline in this discussion.

The year 1922 was important for the lead industry: Clair Patterson was born, and Thomas Midgely of General Motors discovered that tetraethyl lead could be used to control engine knock. The knock was the major limitation on the power of automobile engines, and companies had spent years trying to solve the problem. No wonder the discovery was widely embraced by automobile manufacturers, government officials, and the general public. A representative of Ethyl Gasoline Corporation summarized the mood succinctly when he referred to tetraethyl lead as "a gift of God."

The popularity of leaded gasoline grew rapidly, but the dangers of lead did not go unnoticed. In October 1924, an accident occurred in a New Jersey tetraethyl lead manufacturing plant in which most of the 49 employees at the plant were poisoned. Five workers died, and another 35 suffered severe neurological symptoms such as hallucinations and tremors. There was an immediate public reaction: some members of the health profession and the general public called for the abolishment of leaded gasoline. Company officials, however, blamed the workers for being sloppy in their handling of chemicals at the plant, claiming that both the manufacture and the use of leaded gasoline were safe when the ingredients were handled properly (Rosner and Markowitz, 1985).

The outcry from this and other accidents resulted in the temporary banning of leaded gasoline in several states. By the late 1920's, however, the strength of the lead industry had proven itself victorious and leaded gasoline was legalized throughout the U.S. This situation was to continue for the next forty years. Details of this early period are covered in Chapter 7 by Jerome Nriagu.

What is remarkable about the sequence of events is that nothing was done to stop the production of leaded gasoline after this brief period of controversy despite the knowledge that lead was highly poisonous. To the contrary, leaded gasoline production increased dramatically as the number of automobiles grew. The primary reason is that the lead industry in these years had a virtual monopoly on support of scientific research on lead, and their agenda was protection of the industry rather than pursuit of basic knowledge. As a result, little progress was made for several decades in identifying the true extent of environmental lead contamination or its effects.

Clair Patterson's landmark paper in the journal *Archives of Environmental Health* (1965) on worldwide lead pollution started a new era in attitudes toward lead. This paper, along with other early publications of Patterson, forced the problem of environmental lead out into public debate where industry could no longer control the flow of information (Hays, 1987). To identify the factors that helped Patterson develop his hypothesis in these early papers, let us now consider what was known in the field of geochemistry at the time he began his work.

State of Knowledge of Relevant Concepts in the Field of Geochemistry

Geochemists at the University of Chicago had been working for some time developing methods to determine the age of the earth. In one such method, rocks were ground to a fine powder and tiny amounts of lead and uranium were extracted and analyzed for chemical isotopes. There was a problem, however: the instruments available in the late 1940's were not very sensitive, and thus the measurements could not be made on the oldest rocks. All this changed when Patterson's Ph.D. advisor Harrison Brown, working with Mark Inghram, developed a new highly sensitive mass spectrometer. For the first time, it was possible to measure the isotopic composition of trace amounts of lead and uranium contained in the oldest rocks.

Patterson was encouraged by Harrison Brown to use the technique to determine the isotopic composition of meteorites, from which the age of the meteorites and also the age of the earth could be derived. Using the new instrument, Patterson encountered a problem: there was simply too much lead detected by the sensitive instrument to get an accurate reading. Through a laborious process, working with another graduate student George Tilton, Patterson discovered that careful handling procedures and extensive cleaning of every container, tool, and chemical reagent removed the excessive lead to enable the technique to work. He developed these new procedures while a Ph.D. student and later as a postdoctoral research scientist. Eventually, he was able to establish the age of the earth as 4.5 billion years (Patterson, 1953 and Patterson et al., 1955). It has since been substantiated using other techniques by many scientists.

Thus the state of geochemistry at this time was such that the importance of contamination control was not appreciated. It was not until new instrumentation developed by Brown and Inghram became available that geochemists began to realize the essential role of clean techniques for handling samples when working with small quantities of a substance. As pointed out by Russell Flegal in Chapter 3, luck played a role in Patterson's success: he was at the right place at the right time. Patterson's discovery paved the way for development of clean laboratories for a variety of purposes, from handling environmental samples to manufacturing semiconductors. The detailed story of Patterson's work as a geochemist is told in Chapter 1 by Thomas Church.

State of Knowledge of Lead Pollution Caused by Human Activities

If contamination control was not understood, how well could we measure lead in the environment? The answer is, quite simply, not well at all. The same contamination problem that prevented Patterson from dating the earth for many years also kept scientists, unknowingly, from measuring accurate concentrations of lead. There were plenty of values reported in the scientific literature, but they were mostly wrong. Consider the following examples of evidence that are cited throughout this book:

A sample of seawater was analyzed by Patterson and colleagues using careful contamination control and found to have 14 nanograms of lead per kilogram of water. When the same sample was analyzed by seven other research labs across the U.S., reported values ranged from 50 to 1300 nanograms per kilogram of water, showing severe contamination of the sample by all of the other labs (Patterson and Settle, 1976).

Patterson and co-workers showed that lack of proper techniques during handling of polar ice cores by other scientists resulted in highly contaminated samples. Lead concentrations in the outer sections of the cores where contamination was the most severe exceeded those in the uncontaminated interior by factors up to 100,000 (Ng and Patterson, 1981; Boutron and Patterson, 1983 and 1987).

Settle and Patterson (1980) showed that the muscle of tuna living in the deep ocean contained extremely low concentrations of lead due to the relative lack of anthropogenic lead in this ecosystem. When a sample of this tuna muscle was sent to the U.S. Food and Drug Administration (FDA) for analysis, the FDA reported a value that was 10,000 times larger than the true concentration due to lack of proper contamination control.

Everson and Patterson (1980) showed that lead concentrations in blood plasma were higher in an individual who had excessive occupational exposure to lead compared with those in an individual who did not have this exposure. However, both values were much less than concentrations in blood plasma reported in the literature for a wide range of exposures. Once again, the lack of contamination control in these other studies was apparent.

In unpublished studies, Patterson and others analyzed the lead content of the wood of various trees. The results showed that published lead concentrations in tree wood were erroneous by orders of magnitude. Patterson's values were the only ones that could be used to make estimates of the amount of lead stored in the earth's total biosphere, since trees make up the bulk of the earth's biomass.

The reason for these discrepancies is that other investigators did not appreciate the need for elaborate contamination control procedures. Claude Boutron describes some of these procedures in Chapter 9: extensively acid-washing all containers and tools used to handle a sample, wrapping one's entire body in a large polyethylene bag, and changing acid-washed gloves after each step in working with a sample. These procedures have now been applied to other contaminants. In Chapter 10, William Fitzgerald points out how many of Patterson's original procedures became established for studies of environmental mercury. Analogous to lead, previously published concentrations of mercury were largely erroneous before clean techniques were applied.

It is clear from these examples why scientists and public health officials were unaware of the extent of lead pollution: pre-Pattersonian measurements incorrectly showed that lead concentrations in samples affected by industrial activities were not much greater than they would be under natural conditions. Patterson showed that lead levels under natural conditions were actually much smaller than previously thought, implying that levels in industrial samples once thought to be natural were actually greatly elevated above natural background.

Patterson's findings had important policy implications. For example, the U.S. Food and Drug Administration was unaware of the extent to which lead-soldered cans were contaminating the food contained in them. It appeared that lead concentrations in canned food were similar to those in fresh food before it was placed in the can. Once Patterson's techniques were applied, it became clear that lead levels in fresh food were actually much smaller than in canned food, implying that lead was being leached from the cans into the food. This finding played an important role in developing regulations that prohibited the use of lead-soldered cans for food, as explained in Chapter 1.

All of this information made it obvious that human activities were redistributing lead throughout the environment, significantly increasing human exposure. At least it was obvious to Patterson and his co-workers. The truth is that Patterson faced an uphill battle to convince the scientific community, industry, and government of the implications of his data. The battle lasted for decades. To understand these conflicts, we turn to the second major factor influencing his career.

THE PERSONALITY OF CLAIR PATTERSON

It took an extraordinary individual to confront conventional wisdom that was prevalent around the world. In this section, we consider facets of Patterson's character, especially those that helped him achieve success. We begin by discussing one of his early obstacles, namely the effort to get his paper published in *Archives of Environmental Health* in 1965. The discussion provides a background for several of the chapters that refer to this paper. We then summarize a collection of anecdotal evidence assembled through the years to identify his most important traits.

THE EFFORT TO PUBLISH THE ARCHIVES OF ENVIRONMENTAL HEALTH PAPER

Patterson had moved with Harrison Brown from the University of Chicago to Caltech in 1952 to work as a research scientist. There he continued with geochemical investigations, this time in the ocean, where he inadvertently discovered that the upper levels of ocean water contained far more lead than deeper levels (Tatsumoto and Patterson, 1963; Chow and Patterson, 1966). This marked a turning point in his career. It was at this time, given the new information on lead in the ocean and his earlier experience in dating the earth, that Patterson hypothesized that lead pollution was excessive and worldwide.

For the next several years, Patterson devoted his efforts to putting together a paper that would explain and support his hypothesis. He collected data on production of lead and other heavy metals, and he made estimates of the amount of lead found in the environment during present as well as pre-industrial times. He made similar comparisons for lead in the human body. The results showed that lead concentrations in the environment and in people were elevated orders of magnitude above natural values. These estimates required a great deal of insight, considering the paucity of information available at the time. Looking back, it is noteworthy that many of the estimates remained relatively unchanged as more information unfolded through the years following this early work.

Patterson was invited by the editor of *Archives of Environmental Health*, Katherine Boucot, to submit the results of his work for publication. He obliged by preparing a paper entitled "Contaminated and Natural Lead Environments of Man," which was submitted and sent out for peer review. One of the reviewers was Robert Kehoe of Ethyl Corporation, a key figure in getting public and government acceptance for leaded gasoline. Kehoe's review included the following:

"I recommend the publication of Dr. Patterson's paper, for the reason that it is an interesting view of an important subject.

"It has some exceedingly bad spots in it, as for example the repetitious statement that the data now available in the literature and designated as normal or "normal," (in the sense of current) have been regarded by those who have published such data, to be natural. I, for example, have carefully differentiated between the current sources of the intake of lead and the natural sources, and in 1933 I gathered together much information on the sources of lead, both natural and artificial. There are also several flagrant examples of coming up against facts that do not fit the thesis, and simply discarding them as in error. I shall have to point out some of these things and comment on them at the appropriate time and in the appropriate place. However, it would be useless, I think, to eliminate these faults. I should let the man, with his obvious faults, display these faults. This will keep the record straight, and you will not be neglecting your editorial role, in so doing.

"The inferences as to the natural human body burden of lead are, I think, remarkably naive, and at the same time, from the geochemical viewpoint, very interesting. It is an example of how wrong one can be in his biological postulates and conclusions, when he steps into this field, of which he is so woefully ignorant and so lacking in any concept of the depth of his ignorance, that he is not even cautious in drawing sweeping conclusions. This bespeaks the brash young man, or perhaps not so young, passionate supporter of a cause. In either case hardly the mark of a critical investigator."

The review goes on to state that the paper should not "be swept under the rug. It must be faced and demolished, and therefore I welcome its public appearance."

Despite the fact that Katherine Boucot had invited Patterson to submit the paper, her reply to Kehoe suggests a complete embracing of Kehoe's ideas. Her letter begins:

"Dear Bob:

Long before I ever met you I heard what a tremendous man you are. After I met you I came under your personal spell and, since your field is not intimately related to mine, I never looked at your work critically but always accepted it as top quality, as everyone else did. However, your letter in regard to the Patterson paper is so thoughtful and big in point of view that now, on my own, I realize your full stature and I am extremely proud to have had the privilege of working with you in the past five years."

Patterson's paper was subsequently published, along with highly critical letters to the editor from Kehoe and roughly a dozen other individuals. Representatives of the lead industry also tried unsuccessfully to get Patterson fired from Caltech. And as Patterson's research got underway and began to provide mounting evidence of global contamination by lead, lead industry representatives set up roadblocks at every opportunity. But all this did was to strengthen Patterson's resolve to gather further support for his hypothesis.

What can we learn about Patterson's personality from this episode? Perhaps the table of Pattersonian traits in Chapter 4 by Robert Sharp says it best: he was dedicated, courageous, and forward looking, among other characteristics. As Sharp indicates, he was "intense times 10^3 ". Nothing could deter him, not powerful lead industries nor fellow scientists whose beliefs disagreed with his. Let us now consider what others had to say about Patterson during his life and after his passing that provide additional insight into his personality.

ANECDOTAL EVIDENCE OF A REMARKABLE INDIVIDUAL

Patterson was unconventional in many ways. He refused tenure at Caltech, and he did not accept an award from *Science Citations Index* when one of his papers was cited a record number of times. He also refused to compromise his standards of research in any way, despite forces from industry, government, and fellow scientists. It was a lonely path for many years, especially early in his career.

Several of the chapters discuss personal experiences with Patterson that are quite revealing. For example, Patterson's closest professional colleague was Dorothy Settle who worked with him for nearly 30 years. In Chapter 2, she writes that Patterson "knew exactly how he wanted everything done, he planned every move with meticulous detail, he did countless dry runs on experiments, and moreover he knew what the answer was going to be before any lab work had been done." She also writes about several experiences in working with a man who was brilliant and caring while at the same time driven and uncompromising.

In Chapter 5, Katsuko Saruhashi writes about Patterson's close collaboration with Japanese scientists. In fact, much of Patterson's pioneering early work on lead in the oceans (Tatsumoto and Patterson, 1963), polar ice sheets (Murozumi, Chow, and Patterson, 1969), and remote terrestrial ecosystems (Elias, Hirao, and Patterson, 1982) was done with Japanese collaborators. His great respect for the scientists of this country is evident from his description of the Meteorological Research Institute and Geochemical Laboratory in Japan, cited in Saruhashi's chapter: "This is the best geochemical laboratory, No. 1 in the world." Patterson's close association with the Japanese may reflect a consistency between his strict adherence to contamination control, which was essential to achieve success, and a culture that is well-known for its tradition of strict adherence to authority.

Perhaps more than any other individual, Herbert Needleman shares with Clair Patterson the distinction of fighting the lead industry head-on and achieving success after decades of effort. In Chapter 8, Needleman writes about how Patterson is portrayed as a prophetic figure in Saul Bellow's book *The Dean's December*. The character, known as Sam Beech, is pictured as an eminent scientist warning the world about the dangers of lead. Why the name Sam Beech? It was the closest thing to "sonofabitch," and Needleman points out that not one of Patterson's colleagues is mystified by Bellow's choice of pseudonym.

It took a "Sam Beech" to confront the rest of the scientific community. Indeed, the literature is full of Letters to the Editor and scientific papers challenging Patterson's claims that virtually all previously published values of lead concentrations were erroneously high. Nevertheless, this is a clear case of why one must never assume that scientific opinions with the most "votes" are correct: the majority in this case was wrong. Some of the previous investigators used their incorrect values to reach broad but invalid conclusions. For example, one research group reported erroneously high lead concentrations in glaciers around the world, and concluded that anthropogenic lead emissions from human activities on a global scale were less than 10% of the total lead emissions. This work was promptly put down by Patterson who showed that when correct concentrations were used, it could be demonstrated that lead from human activities overwhelmed natural lead emissions (Patterson, 1983).

At the Caltech memorial service for Clair Patterson on March 8, 1996, Russell Flegal made an interesting observation. Having spent three long days telephoning colleagues with news of Patterson's death, he commented that if there were an entry in the Guiness Book of World Records for the size of Scientific Families, Clair Patterson would set the record. Indeed, he touched the professional lives of so many scientists that it would be impossible to count all of them.

Patterson also touched the lives of many people who were not scientists. Many of the talks at the memorial service showed that he had a deep concern for humankind. His son Cameron observed that when his father saw children, he viewed them as future scientists, future artists, and members of other important professions. Cameron noted that this was his father's reason for being motivated to protect the environment. His nephew Paul, currently a Professor of Biology at Caltech, commented on an experience when he was 7 or 8 years old. His uncle drew some boxes and arrows on a sheet of paper and told Paul that these were plans for the atomic bomb -- and must be kept secret. Paul noted that this was his uncle's way of initiating him into the community of scientists.

Todd Hinckley noted that Patterson made him itch in so many places that it was impossible to scratch all of them. As pointed out by Hinckley, Patterson was far ahead of the rest of us, and it may take some time before we extract all of the useful information from his writings.

At his retirement party in December 1992, a number of stories were told. George Tilton recalled Patterson's impatience at some conferences. He and Patterson were together at a conference where a traffic stoplight in the front of the room was used to signal the end of each talk. As Patterson was talking, the light changed from green to yellow, then red. Patterson merely walked over to the stoplight and unscrewed the bulb! The session chair sat in amazement as Patterson took another ten minutes to finish his talk.

Ken Bruland noted that he and Patterson were on a ship heading out to collect seawater samples when a storm hit them. Patterson became white as a ghost with seasickness. But his real agony was worrying that they might terminate the trip on account of his sickness and head back for shore. "Please, Ken," he exclaimed, "Promise me one thing: don't let them turn the ship around until we've got the samples!" They got the samples. When Patterson, still quite seasick, saw an ambulance waiting for him as they pulled into the dock, he was furious. "Get the hell out of here," he yelled. "We've got samples to analyze!"

Bob Duce told about getting Patterson involved in the SEAREX Program on the Pacific Island of American Samoa. A tall tower had been set up to enable air sampling well away from the ground, and one of the filters on top was to be used to analyze for lead. It was difficult going up and down 175 steps on the tower each time a visit to the sampling platform was necessary. However, it was especially difficult at night since Patterson would not allow any of the tower lights to be turned on for fear that an insect, attracted by the lights, would be sucked into his filter sampler. Duce also recalled the one instance when he obtained a number for airborne lead concentration that was lower than Patterson's -- and then discovered the value was for a damaged filter that had a hole in it!

Tom Church related his early experiences with Patterson. Their first acquaintance was made at MIT while Patterson was on sabbatical there. Church recalled that Patterson was wearing two different colored socks, and expounded about how children in inner-city Boston were being exposed to enormous amounts of lead. Church later went to Caltech to work briefly in Patterson's lab. When he tried to enter the lab for the first time, he was stopped: "We'll begin at ground zero!" he recalled Patterson saying. And indeed they did, by getting down on their hands and knees together and cleaning the floor with Kimwipes, square centimeter by square centimeter.

Robert Elias noted that Patterson, as a Celt, had behavior that was consistent with the three well-known characteristics of Celtic culture. First, there is no conventional written language, so Celts must have all their knowledge in memory. Second, Celts are lucid oral communicators and great teachers. And finally, Celts will fight to the finish -- and then march off the battlefield with the heads of their opponents on their spears!

Karl Turekian recalled that he reviewed the 1980 National Academy of Sciences report on lead in which Patterson had written a chapter entitled "An Alternative Perspective -- Lead in the Human Environment: Origin, Extent, and Significance." Turekian strongly backed the viewpoint in this chapter. When Patterson heard about this review, he was so elated that he immediately phoned Turekian to say how pleased he was -- not realizing that it was 2:00 o'clock in the morning!

Bruce Doe told about a conversation as he was touring Patterson's lab in the 1960's. He asked whether there was any hazard associated with the still for purifying perchloric acid in the corner of the lab. Patterson's response: "No problem. If that still explodes, the entire building will be gone and you won't feel a thing. And if you want to ask about the ether still over there, it's also no problem. If that still goes, the building will be completely engulfed in flame before you have a second thought."

Indeed, Patterson believed that a scientist must take whatever risks are necessary to pursue scientific truth. He once attended a meeting of glaciologists where choice of a lubricant fluid for drilling ice cores was being debated. One fluid was relatively pure but highly toxic, while another had slight amounts of contaminants but was much less toxic. After listening to the debate for a while, Patterson got up and announced his preference for the purer fluid: "There is no real choice here. As a scientist, you must give your life to science." This was one situation where Patterson could not convince the group, however.

Nobel Laureate Roger Sperry wrote the following at the time of Patterson's retirement in 1992:

"Occasional encounters with Clair Patterson stand out among the more memorable aspects of my earlier Caltech years when I used to regularly walk the campus fighting the "use-it-or-lose-it" war with PLS. I had come to know Clair initially through noon-hour activities at the Athletic Center where I'd swim or check out a basketball and Clair would often be running around the track -- usually the only one going in the opposite direction from everyone else (to favor his weak side). An encounter with Clair usually meant a chat about some esoteric, far-out thought he happened to be musing on including, not only the hidden evils of lead, but most anything, even the strange mind of the bisected-brain about the implications of which Clair characteristically had more penetrating insights and interest than many closer colleagues. It's hard to believe that Clair might really be planning now to retire."

Scientific research is sometimes depicted as a tortuous path subject to human weaknesses, for example as told by Thomas Kuhn in his book *The Structure of Scientific Revolutions* (1962) cited in Chapter 8 by Needleman. Patterson's career clearly illustrates the struggles of modern research as well as the surprising turns that research can take. Although he started out as a geochemist, the pursuit of knowledge of lead pollution forced Patterson to consider topics such as the history of mining and smelting of metals, transport and deposition of pollutants from the atmosphere to the earth's surface, the behavior of nutrients as well as pollutants in ecosystems, and human exposure and toxicology of lead. At the time of his death, Patterson was formulating new concepts on the way the human mind had evolved over the past thousands of years, and was preparing a book intended for young scientists on future research on the philosophy of science. This is discussed further by Church in Chapter 1.

Patterson received numerous recognitions over the years. These occurred despite the fact that he often stated his motivation for the work was science itself, and tried to avoid recognition for his work. Some of the more important recognitions include the following:

The asteroid 2511-1980 LM, a 120 kilometer meteoritic body, was named after Patterson.

A peak in the Queen Maude Mountains of Antarctica (85° 44' South, 155° 59' West) was named after Patterson.

Patterson won the following awards during his life: the Tyler Prize for Environmental Achievement (1995), the Professional Achievement Award of the University of Chicago (1981), the Goldschmidt Medal of the Geochemical Society of America (1980), an Honorary Doctorate from the University of Paris VI (1975), an Honorary Doctorate from Grinnell College (1975), and the J. Lawrence Smith Medal of the National Academy of Sciences (1973).

The success of this striking career can be illustrated in many ways. But perhaps the most noteworthy illustration is evidence of decreased production of leaded gasoline worldwide. This is a victory shared with Clair Patterson by several of the authors of this book, and others, who devoted their careers to understanding the environmental behavior and effects of lead. The table following this Introduction lists the lead contents of gasoline in countries around the world as of 1995 (Mauss, 1996). It is clear that tremendous progress has been made in reducing lead emissions from motor vehicles: this stands out as one of the most notable environmental achievements of the twentieth century.

Clair Patterson's life is a model of perseverance and unwavering dedication to pursuit of the truth through scientific research. The pages that follow present observations and interpretations of his life through the eyes of ten different individuals. It is hoped that by considering these different perspectives, the reader will be able to form his or her own opinions about this remarkable individual. And by studying the lives of great people such as Clair Patterson, perhaps each of us can gain a better understanding of both scientific progress and human nature.

Structure of the Book

This book is divided into two sections, each containing five chapters. Section I is entitled "Clair Patterson: The Person" and includes material on Patterson's life history, personality, and experiences. Section II is entitled "Professional Contributions of Clair Patterson and Colleagues." This section includes a summary of his scientific findings and those of individuals who knew him and worked with him.

Section I

Chapter 1 is a chronology of Patterson's life, essentially told by Patterson himself through a series of interviews by colleague Thomas Church. The chapter begins with early childhood and traces his experiences in college and graduate school, through his career at Caltech, and ultimately to his last research effort at the time of his death.

In Chapter 2, Dorothy Settle recalls some of her experiences with Patterson during her employment as a research scientist in his laboratory for nearly thirty years. The chapter includes several humorous as well as sobering encounters at Caltech and during their sampling trips.

In Chapter 3, Russell Flegal, whose Ph.D. work was supervised by Patterson, calls attention to Patterson's personality, his support from family and colleagues, and luck. The chapter focuses on the last of these factors by summarizing many of the circumstances where luck played out in Patterson's favor.

Chapter 4 is a discussion of Patterson's place in the Caltech environment as told by Robert Sharp, former Director of the Division of Geological Sciences there. Beginning with Patterson's arrival at Caltech and construction of his unique clean laboratories, the chapter relates several stories about his relations with others, especially some of the more humorous interactions.

Chapter 5 presents some remembrances of Katsuko Saruhashi, a colleague of Patterson from the Meteorological Research Institute and Geochemical Laboratory in Japan. The chapter mentions Patterson's visits to Japan as well as his work with Japanese scientists in the U.S.

Section II

Chapter 6 by Cliff Davidson, who worked in Patterson's laboratory for part of his Ph.D. work, summarizes Patterson's scientific contributions in each of the three phases of his career. These include development of his original hypothesis on the ubiquitous nature of lead, experiments to test the hypothesis, and use of the data to explore the significance and implications of the work.

In Chapter 7, colleague Jerome Nriagu describes the events leading to widespread use of leaded gasoline in the 1920's. The chapter also discusses Patterson's role in the

1960's in working to change public and government attitudes about lead in gasoline that eventually resulted in removal of tetraethyl lead as a gasoline additive.

Chapter 8 by colleague Herbert Needleman chronicles how Patterson's early hypotheses and findings were attacked by industry and the scientific community, only to be embraced by scientists decades later. Needleman then shares his experiences in attempting to gain acceptance of his own findings on the health effects of lead in children, drawing parallelisms with Patterson's life-long efforts.

Chapter 9 by colleague Claude Boutron presents data for lead and other metals in polar snow analyzed at his own laboratory in France, and compares the data with Patterson's values to demonstrate the difficulty of achieving adequate contamination control. Results from his later collaborative research with Patterson are discussed, along with details of the experimental procedures including photographs.

Finally, Chapter 10 by colleague William Fitzgerald describes pioneering work in measuring mercury in environmental samples, with procedures modeled after Patterson's original techniques developed for lead. The chapter includes experiences working with Patterson as well as data for mercury in surface waters.

Cliff Davidson

Leaded Gasoline in Countries around the World, 1994-1995

(Data compiled by Associated Octel, Ltd., England; provided by Mauss, 1996)

	Total vehicle consumption 10^9 liters	Lead content of leaded gasoline grams/liter	Leaded gasoline share percent
North America			
Canada	32	--	0
United States	430	--	0
Mexico	26	0.06-0.28	54
Central America			
Belize	~0.04	0.84	100
Costa Rica	~0.41	0.2	67
El Salvador	--	0.7	<100
Guatemala	0.41	--	0
Honduras	~0.2	0.84	100
Nicaragua	--	0.31	?
Panama	~0.3	0.82	99
Carribean			
Antigua	0.02	--	0
Bahamas	--	0.84	38
Barbados	--	0.84	<100
Cuba	--	0.84	96
Haiti	--	0.84	100
Puerto Rico	2.9	--	0
Trinidad/Tobago	0.36	0.4	100
South America			
Argentina	5.3	0.15	62
Bolivia	--	0.42	100
Brazil	17.0	--	0
Chile	1.7	0.34	82
Colombia	6.4	--	0
Ecuador	~1.8	0.32	79
Paraguay	--	0.84	100
Peru	1.7	0.84	--
Surinam	0.06	--	0
Uruguay	--	0.3	98
Venezuela	9.9	0.37	100

LEADED GASOLINE IN COUNTRIES AROUND THE WORLD, 1994-1995 (CONTINUED)

(Data compiled by Associated Octel, Ltd., England; provided by Mauss, 1996)

	Total vehicle consumption 10^9 liters	Lead content of leaded gasoline grams/liter	Leaded gasoline share percent
Africa			
Algeria	2.6	0.5-0.6	100
Botswana	--	0.44	<100
Egypt	3.0	0.4	<100
Ethiopia	--	--	<100
Libya	2.0	0.4-0.8	100
Madeira	~1.0	0.4	85
Morocco	--	0.5	<100
Nigeria	--	0.66	100
Reunion Islands	~1.0	0.15	85
South Africa	6.4	0.3-0.4	100
Middle East			
Israel	2.1	0.05-0.15	85-90
Jordan	--	0.09-0.4	<100
Palestinian Authority	--	0.15	85-100
Saudia Arabia	9.3	0.4	<100
Syria	1.5	0.24	100
Far East/Oceania			
Australia	17.0	0.15-0.40	47
China	27.0	0.08-0.25	42
Hong Kong	0.35	0.15	29
India	4.8	0.15-0.42	100
Indonesia	--	0.45	100
Japan	44.0	--	0
Malaysia	3.8	0.15	?
New Zealand	2.5	0.18-0.34	54
Singapore	0.58	0.15	39
South Korea	3.8	--	0
Sri Lanka	0.23	<0.4	100
Taiwan	4.9	0.05-0.08	34
Thailand	3.7	0.15	52

LEADED GASOLINE IN COUNTRIES AROUND THE WORLD, 1994-1995 (CONTINUED)

(Data compiled by Associated Octel, Ltd., England; provided by Mauss, 1996)

	Total vehicle consumption 10^9 liters	Lead content of leaded gasoline grams/liter	Leaded gasoline share percent
Western Europe			
Austria	3.5	--	0
Belgium	3.7	0.15	35
Denmark	2.2	--	0
Finland	2.7	0.15	<30
France	27.0	0.15	54
Germany	42.0	0.15	8
Greece	3.3	0.4	72
Italy	19.0	0.15	66
Netherlands	4.7	0.15	25
Portugal	1.9	0.10-0.4	70
Spain	11.0	0.15	78
Sweden	5.7	0.15	<1
Switzerland	5	0.15	21
Turkey	4.4	0.15-0.4	97
United Kingdom	33.0	0.15	42
Eastern Europe and Russia			
Armenia	--	--	100
Bulgaria	--	0.15	<100
Czech Republic	<2.3	0.10-0.14	62
Hungary	2.1	0.15	--
Iceland	--	0.15	16
Poland	3.7	0.15-0.30	77
Romania	28	0.3-0.6	--
Russia	~100	0.17-0.37	--
Slovakia	<2.3	0.15	19

REFERENCES

Boutron, C.F. and Patterson, C.C., The occurrence of lead in Antarctic recent snow, firn deposited over the last two centuries and prehistoric ice. *Geochimica et Cosmochimica Acta*, Vol. 47, pp. 1355-1368 (1983).

Boutron, C.F. and Patterson, C.C., Relative levels of natural and anthropogenic lead in recent Antarctic snow. *Journal of Geophysical Research*, Vol. 92, pp. 8454-8464 (1987).

Chow, T.J. and Patterson, C.C., Concentration profiles of barium and lead in Atlantic waters off Bermuda. *Earth and Planetary Science Letters*, Vol. 1, pp. 397-400 (1966).

Elias, R.W., Hirao, Y., and Patterson, C.C., The circumvention of natural purification of calcium along nutrient pathways by atmospheric inputs of industrial lead. *Geochimica et Cosmochimica Acta*, Vol. 46, pp. 2561-2580 (1982).

Everson, J. and Patterson, C.C., "Ultra-clean" isotope dilution/mass spectrometric analyses for lead in human blood plasma indicate that most reported values are artificially high. *Clinical Chemistry*, Vol. 26, pp. 1603-1607 (1980).

Hays, S.P., Beauty, Health, and Permanence: Environmental Politics in the United States, 1955-1985, Cambridge University Press, Cambridge, England, p. 348 (1987).

Kuhn, T., *The Structure of Scientific Revolutions*, University of Chicago Press, Chicago, Illinois (1962).

Mauss, E., Status of the global campaign to phase out leaded gasoline. Presentation at the Sixth International Conference of the Israeli Society for Ecology and Environmental Quality Sciences, Jerusalem, Israel, June 30-July 4, 1996.

Murozumi, M., Chow, T.J., and Patterson, C.C., Chemical concentrations of pollutant lead aerosols, terrestrial dusts and sea salts in Greenland and Antarctic snow strata. *Geochimica et Cosmochimica Acta*, Vol. 33, pp. 1247-1294 (1969).

Ng, A. and Patterson, C.C., Natural concentrations of lead in ancient Arctic and Antarctic ice. *Geochimica et Cosmochimica Acta*, Vol. 45, pp. 2109-2121 (1981).

Pacyna, J.M., *Atmospheric trace elements from natural and anthropogenic sources, in Toxic Metals in the Atmosphere*, J.O. Nriagu and C.I. Davidson, (Eds.), Vol. 17 in the Wiley Series Advances in Environmental Science and Technology, John Wiley and Sons, New York, pp. 33-52 (1986).

Patterson, C.C., The isotopic composition of meteoritic, basaltic and oceanic leads, and the age of the earth. Report by the Subcommittee on Nuclear Processes in Geological Settings, National Academy of Sciences/National Research Council, Washington, D.C., pp. 36-40 (1953).

Patterson, C.C., Contaminated and natural lead environments of man. *Archives of Environmental Health*, Vol. 11, pp. 344-360 (1965).

Patterson, C.C., Criticism of "Flow of metals into the global atmosphere." *Geochimica et Cosmochimica Acta*, Vol. 47, pp. 1163-1168 (1983).

Patterson, C.C., Tilton, G., and Inghram, M., Age of the earth. *Science*, Vol. 121, pp. 69-75 (1955).

Patterson, C.C. and Settle, D.M. (1976) The reduction of order of magnitude errors in lead analyses of biological materials and natural waters by evaluating and controlling the extent and sources of industrial lead contamination introduced during sample collecting, handling, and analysis. In*: Accuracy in Trace Analysis: Sampling, Sample Handling, and Analysis*, P. LaFleur (Ed.), National Bureau of Standards Special Publication 422, pp. 321-351 (1976).

Rosner, D., and Markowitz, G., A gift of God?: The public health controversy over leaded gasoline during the 1920's. *American Journal of Public Health*, Vol. 75, pp. 344-352 (1985).

Settle, D.M. and Patterson, C.C., Lead in albacore: guide to lead pollution in Americans. *Science*, Vol. 207, pp. 1167-1176 (1980).

Smith, K.R., Fesharaki, F. and Holdren, J.P. (Eds.), *Earth and the Future: Essays in Honor of Harrison Brown*. Westview Press, Boulder and London (1986).

Tatsumoto, M. and Patterson, C.C., Concentrations of lead in some Atlantic and Mediterranean waters and in snow. *Nature*, Vol. 199, pages 350-352 (1963).

U.S. Environmental Protection Agency, Air Quality Criteria for Lead, Report EPA-600/8-77-017, p. 5-3 (1977).

PATTERSON'S ORIGINAL CARTOONS

One of Patterson's many talents was drawing cartoons. We are fortunate to have examples of some of the cartoons he drew while in Antarctica. They are presented here, along with accompanying text written by Patterson himself describing the working conditions on this trip.

"In 1966, seven graduate students and I spent three months of the summer field season in the Antarctic collecting snow samples for chemical analysis. The purpose was to see whether the increasing industrial contamination of the atmosphere in recent years would be reflected as progressively dirtier layers of snow in a place where the snow had not melted each year or slid down inclines and mixed up. That is, preserved snow layers were being used as indicators of atmospheric conditions of the past. The dirt being studied in the snow was lead, and it occurred in such small concentrations that we needed a hundred gallons of snow for an analysis. This meant that we had to mine the snow out. Logistically our operation was complicated and tedious. Most of the work was done near Byrd Station, a permanent installation in the middle of a vast plateau of mile-thick ice within the Antarctic interior.

Four men in the project took time out from their labor of digging a three hundred foot shaft near Byrd to make a sled traverse with me to a virgin area one hundred and thirty miles upwind of Byrd for the purpose of collecting recent samples of snow that would not have been dirtied by vehicular activity around Byrd. During most of the two weeks required for this operation we were camped at a single location and enjoyed one warm spot, the Sno Cat tractor, which served as the locomotive for our sled train. This vehicle contained a stove, a kitchen, and tables with chairs. Conditions were crowded, and most of us lived in tents pitched in the snow alongside the sleds, and only ate our meals and dried our clothes in the Sno Cat. A routine was developed whereby I, as fearless leader, arose first in the morning and went to the Sno Cat, dressed, roused the cook, and then the others. To keep out of the way of the men while they were dressing and eating in the main tractor body, I climbed forward and sat in the observer's seat next to the engine, and to while away time, drew cartoons.

So each morning during breakfast, I drew one or more cartoons for the edification of my mates. My only training in the drawing arts was gained in a high school mechanical drawing class, and the cartoons turned out to be rather primitive. Instead of serving as morning comics, these efforts elicited only sour grunts or yawning mumbles. This was not a reflection of sophisticated tastes. On the contrary, the men in my party were graduate students in the physical sciences, which persons, everyone knows, have pigeonhole intellects and therefore enjoy simple pleasures. It is possible, however, that their attitudes did not express a fair appraisal of my graphic art, but instead resulted from the unfortunate circumstances that we had arrived at, and were working in, an area of heavy snow accumulation in a warm climate, that our supposed gentle lark had turned into another grinding salt mine ordeal.

We expected light accumulation in a cold environment, and had planned to dig a shallow trench in soft material so that we could fill a dozen barrels with snow scooped from successive layers accumulated during the last fifteen years at that spot. Upon arrival, however, stratigraphic evidence showed that we must dig down fifty feet into ice firn to reach a layer deposited fifteen years ago. We were forced to knuckle down to the pick, the shovel, and the man-haul, and dig an inclined shaft one hundred feet long to provide access to the snow layers that were to be sampled. One member of the party, in bitter contemplation, calculated that we hoisted nearly one thousand banana boat loads of ice up and out of that slanting hell-hole.

Unabashed, therefore, I offer these drawings for perusal, with a reminder that their primitiveness is compensated for by their authenticity. Some of the drawings were half-sketched ideas that were finished later. In others the grease spots and food smudges were so bad, they were re-traced. The spoof about the cold food is not really a dig against our cook-mechanic. On the contrary, his determined inventiveness provided, first, an oil stove that really heated, second a gigantic snow melter that gave us prodigious amounts of water, and third, warm food in dishes that re-appeared magically clean each time we stumbled back into the Sno Cat, cold and exhausted after work.

A distinction is made in these cartoons between two kinds of men who worked down there -- members of the United States Navy and civilian scientists. For every scientist carrying out an investigation in the United States Antarctic Research Program (USARP), there were a number of Navy men, largely Seabees and Airdales, providing logistical support. These two different groups of men did not speak the same language, but they worked closely together. Intergroup friction decreased in proportion to the decrease in responsibility attached to the individuals involved, and this happy circumstance permitted a great deal of amicability, for there were not many senior scientists nor officers down there.

Within our own party personal problems could have been more difficult. The chill mists of a Scottish heritage have tinged my outlook with gloomy introspection, but the irrepressible good cheer and enthusiasm of the rest of the men usually overrode this, which made the expedition stimulating and satisfying to everyone."

No, the one without a beard is a male too. He isn't permitted to look ridiculous here because he's already made a fool of himself some other place.

Do I put in the log how we got here or just where we are?

All that melted snow - incredible!

"....while you were enduring privation and hardship at the ends of the earth, the theoretical basis for your investigations was shown to be erroneous. It is now apparent that data of greater significance can be collected by trained monkeys working part-time in comfortable, air-conditioned laboratories. You are hereby notified that further support of your investigations will be terminated as of...."

It must be some sort of hot weather lubricant.

We may be thin on defense, but we're tigers on boosting Seabee morale.

OK, I'll double the length of the poles
and halve the cost of the accumulation data.

Are we still at the same place we were
yesterday, or have we moved again?

No, I'm headed for town. You're headed for the lab!

Reveille! Reveille! Wind 25 knots.
Sky overcast. Temp. minus twenty.
Coffee iced. Pancakes frozen. Bacon half-thawed.

I understand everything but one part -
who ordered you to do this?

SECTION 1

CLAIR PATTERSON: THE PERSON

CHAPTER 1

FROM METEORITES TO MAN: THE PATTERSON GEOCHEMICAL HERITAGE

THOMAS M. CHURCH
College of Marine Studies
University of Delaware
Newark, Delaware 19716-3501

ABSTRACT

Clair Cameron "Pat" Patterson (1922-1995) had a career in geochemistry which over 40 years witnessed a remarkable evolution from the dawn of nuclear chemistry to the cutting edge of mankind's place in the global environment. The measure of scientific success for Pat was not so much in his many individual accomplishments as it was in guiding the evolution of a field. The field in this case is geochemistry, and the evolution is a matter of not just the intellectual gifts Pat contributed, but how he, more than most, affected the peer heritage in the field of trace element geochemistry. The Patterson geochemical heritage is traced from his boyhood roots to the Chicago graduate days born out of the torments of the Manhattan project, to environmental concerns about trace element pollution of the globe, and finally to man's functioning in the Global Commons as it may be affecting the future of the human mind.

1. BOYHOOD ROOTS

The origins of one the country's great geochemists were humble indeed. Clair Cameron (Pat) Patterson was born in 1922 in the small rural town of Mitchellville, Iowa. As is the case today, this part of the American heartland had a low population density. However, in those days of little automobile traffic and no TV, a rewarding social relationship, by fellowship if not by blood, was established within a sort of tribal society that grew up and lived together. The boy Patterson (Figure 1) was among a close-knit group of youngsters who at play learned the intimate natures of plants, animals and rocks, both wild and domestic, as they camped, fished, swam, and hunted in the bottomlands of the nearby Skunk River. Even at work this kinship was nurtured as they observed first-hand the milking of cows, the feeding of steers and butchering of hogs, and the planting and

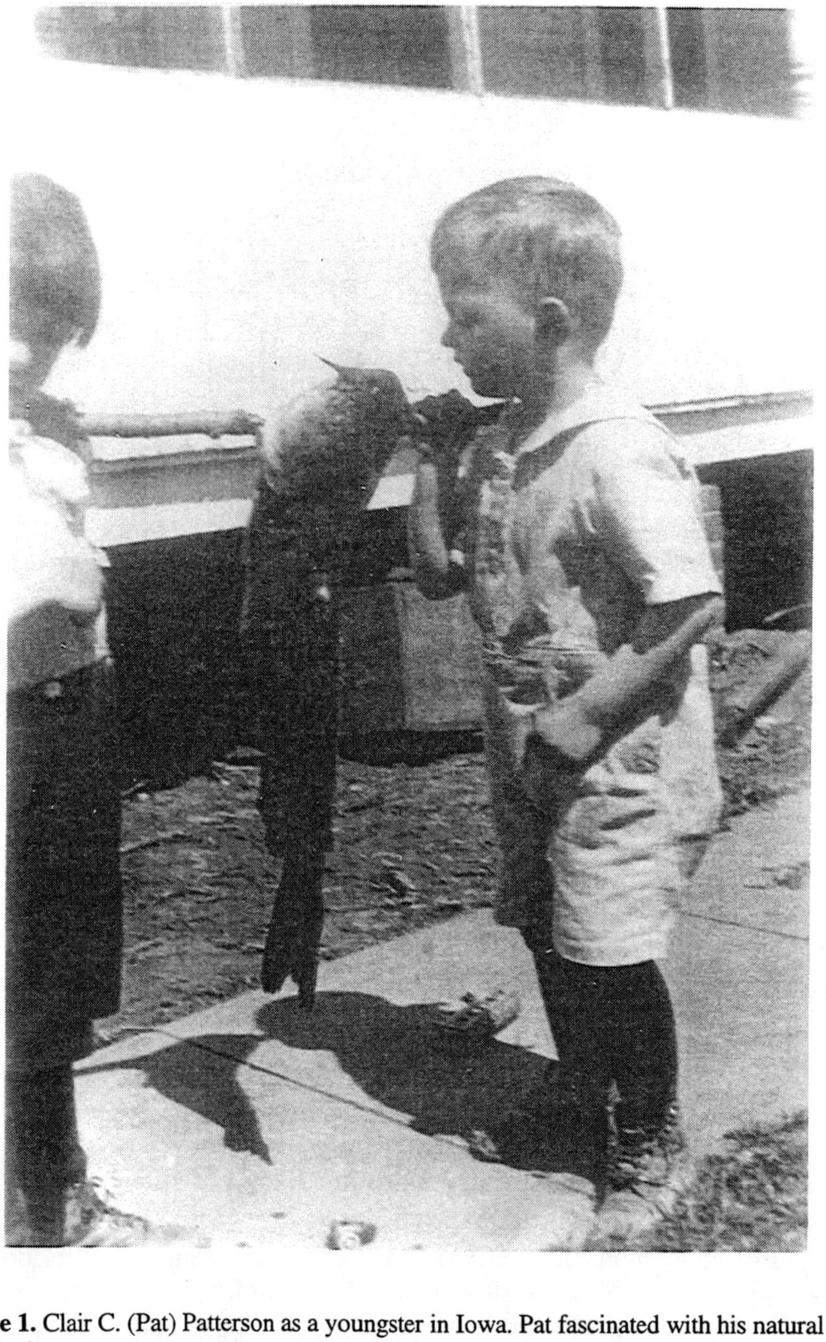

Figure 1. Clair C. (Pat) Patterson as a youngster in Iowa. Pat fascinated with his natural rural environment, in this case a large catfish.

harvesting of crops in nearby farms of friends. So Pat became fascinated with this natural world, and began collecting and puzzling over bones and Indian artifacts.

Times were hard as the depression set in and farmers resorted to such things as burning their corn for fuel. His father was luckier than most. As a rural mail carrier, he had a steady income. Both mother and father and many relatives were college educated in the sciences and other subjects.

Pat's early exposure to experimental science actually started at age 12 when he was a high school freshman. His uncle, who had been a chemistry student at Iowa State College, gave him some chemistry texts and laboratory manuals. His mother and the school superintendent obtained chemicals and labware for him. Pat, armed with these, set up a laboratory in the basement of his home where he carried out experiments in qualitative analyses and taught himself to develop an intuitive appreciation for the chemical rhythms of the periodic table. Pat was guided by his math teacher, who was also the school superintendent for a small high school. The school had less than 100 students, and all of them had spent their lives together. He was a top ten finalist among high school algebra and geometry students in state-wide yearly tests given by the University of Iowa.

Pat's superiority in science and math was unobtrusive to his classmates. They tolerated his particular differences as being a common manifestation by all members of the small high school tribe to exhibit differences among themselves of one kind or another. Nevertheless he was a young, selective, self-initiator who refused to fit the mold by the time he was a 16 year-old high school graduate.

2. College Years

Pat entered Grinnell College in 1939, where his rebellious nature was guided toward positive ends by caring mentors of a small, gifted faculty. He was also helped in his math background by one of his tribal boyhood friends and classmates, Joe Dykstra. Although he had excelled in math in high school, he rebelled against homework time spent on math and received a low grade in differential equations as a consequence.

At Grinnell, he met and befriended Lorna McCleary while they were colleagues in the laboratory and classroom during their four years of majoring in chemistry. Their mothers had grown up together in the same small Iowa town before moving away as adults. As a result of the mutual cooperation between Pat and Lorna in studies and laboratories, Lorna received top grades in chemistry. Pat, a scornful rebel in class behavior, fared less well in formal grades. However, his scientific thinking in chemistry and physics was greatly enhanced and broadened through influences from his chemistry professors Leo Sherman and Bill Oelke, and physics professor Grant Gale.

On a nearby shelf in one of his classrooms were some meteorite fragments given to Grinnell College by Dr. Nininger, who was well-known at that time for his work in recovering the remains of fallen meteorites. The fragments had been accepted by Grinnell as partial payment of tuition for his daughter Doris, who was a classmate of Pat's in chemistry. Among those fragments was an iron meteorite that came from Canyon Diablo Meteor crater in Arizona, which supplied the piece of iron meteorite that Pat later used to determine the age of the earth.

In the small labs at Grinnell, students were largely self-taught. Through one-on-one guidance from their professors, they invented a large portion of their own experiments. In one unfortunate instance Pat's creative urges led him to concoct some experiments in third year organic chemistry which inadvertently blew up one end of the lab. Although Professor Sherman severely reprimanded Pat for this, he nevertheless later recommended him highly to Graduate Schools and encouraged him to enter graduate study.

The Dean of Men, Shelton Beatty, countenanced Pat's rebellious behavior throughout his years at Grinnell. Then, a few months before he was to graduate at Grinnell, and when he had been accepted for graduate study and was set to enter the University of Iowa, the Dean called Pat into his office. The list of sins he had forced Grinnell College to endure over the years was read to him, and he was expelled for two weeks from Grinnell as a clear message that those sins had been noted, were unacceptable, and should not be repeated. The Dean said he acted then instead of earlier because he believed Pat might have reacted by abandoning college, so he waited until Pat was about to graduate, committed to graduate school, and the message could have a positive impact without permanent harm.

In 1943 during the middle of World War II, Pat continued on to graduate school at the University of Iowa. Lorna worked in a defense plant in Davenport, Iowa and in research laboratories in Appleton, Wisconsin and Iowa City. They were married at the time he completed an MS degree in Physical Chemistry. His MS work was completed in less than nine months under the tutelage of George Glockler, a famous molecular spectroscopist who later transferred to Duke University. Pat spent part of his time as a teaching assistant in classes for army personnel, medical students and nurses. For his thesis, Pat investigated extents of depolarization of certain modes of molecular vibrations in halogenated organics using Raman spectroscopy. In those times, depolarization factors in Raman spectra helped define the character of the vibration assigned to a given spectral frequency. These data were tediously developed on photographic plates that were read by an optical densitometer that reported its results through a rotating galvanometer mounted on a vertical wire that stretched from ceiling to floor. Current strengths were indicated as reflected light marks from the galvanometer mirror shining on a circular scale three feet in diameter.

As part of the war effort, Pat was transferred in 1944 to the Metallurgical Laboratories at the University of Chicago, where he worked as an atomic spectroscopist in an analytical laboratory dealing with the Manhattan Project. Here he carried out operations related to studies of trans-uranic and fission product elements, which were in turn related to development of knowledge concerning nuclear fission. Using emission spectroscopy, he started to analyze minute quantities of fission products as part of the Manhattan Project. At that time, this was the method of choice to describe the new trans-uranic elemental properties being used and produced for the first atomic bombs. These were the earliest experiments of nuclear fission, and amidst much experimental fervor, the atomic bomb neared completion.

3. THE WAR AND THE BOMB

After a few months in the Chicago Metallurgical Laboratories, Pat despaired at being the only young man there. Out of a sense of duty, he returned to Iowa to have his draft board enlist him, and Lorna enlisted in the Waves. However, after only two days, they were told that the Iowa Draft Board had no jurisdiction over Pat as long as he was serving as a civilian in the Chicago lab. When Pat and Lorna returned to Chicago, a Fifth Army Colonel interviewed them and asked if they might be happier at the Oak Ridge National Laboratory in Tennessee, where many young men were serving on the Manhattan Project. Pat chose the latter, a virtual "scientific concentration camp" where scientists worked around the clock. Pat and Lorna went to the Tennessee Eastman Y-12 segment of the Oak Ridge, Tennessee plant. Here, ^{235}U atomic bomb fuel was produced by huge mass spectrometers, with uranium hexafluoride sources, milliamp ion currents, and collector boxes of slotted graphite. These huge machines, called "calutrons", had coils wound with most of the US Government's supply of silver taken from vaults of the US Treasury. The vacuum beam path of each machine was in a fifteen by twenty foot plane isolated in separate steel boxes placed some ten feet apart. But those beams were wrapped around magnets that were connected in a common magnetic loop about the size of the perimeter of a football field. These machines enriched ^{235}U from a small fractional abundance in common uranium to high purity in a series of repeated passes through them. By such means of repeated chemical processing, successively purified material was collected into source material between each stage of enrichment.

At first, Pat and Lorna worked in analytical labs checking the chemical purities of these materials. Later, Pat switched to maintaining small mass spectrometers that were used in analytical laboratories to check isotopic purities of enriched bomb fuel at various stages of production in the calutrons. Thus it was at this time he learned the principles and operation of mass spectrometry which was to be the analytical technique he used for most of his career. Here, Pat again encountered Joe Dykstra from Grinnell, who was working in the Carbon and Carbide K-10 section gaseous diffusion enrichment segment of the Oak Ridge bomb fuel production plant. Joe spent his entire career at Oak Ridge National Labs and today still consults on the use of ^{235}U both in the US and in the former Soviet Union.

Although scientists involved in the Manhattan Project knew of the destructive horror of the atomic bomb, they had worked without remorse in a cooperative effort "to save democracy from the horrors of fascist slavery." Some also believed that the lives of American soldiers, probably saved by use of the bomb, justified the scientists' efforts and avoided an invasion of Japan. Nevertheless many of the scientists felt both responsibility and remorse for the incineration of hundreds of thousands of Japanese civilians and the endowment of humankind with the most evil war weapon yet devised. However, many of these scientists returned to their universities brimming with bold new concepts in peaceful uses of nuclear chemistry. The group returning to academia at the University of Chicago included such individuals as William Libby, who pioneered ^{14}C dating, Harold Urey, who pioneered paleo-temperature determination, and Harrison

Brown, Pat's thesis advisor. It was Brown who was to pioneer nuclear geochemical concepts dealing with the origins of stars and the solar system, and U/Pb geochronology in common rocks.

4. Time for Lead

At Johns Hopkins, Brown had earned his doctorate by working on cobalt isotopes in the context of nuclear formation of elements in stars. There he had helped pioneer the use of mass spectrometry in geo- and cosmochemical studies in a manner similar to that of Alfred Nier. Even before the war, Nier had established the scientific basis for the quantitative relationships between various lead isotope abundances and rates of decay and abundances of different isotopes of uranium. One consequence of these studies by Nier was a concept developed by Brown of how the age of the earth could be determined. This age was an unknown parameter in a mathematical relationship of uranium isotope abundances and decay parameters, and of lead isotope abundances, both in the earth today and at the time it was first formed. All parameters in this mathematical expression that could yield the age of the earth were known except the primordial isotopic composition of lead in the solar system at the time of its formation. If the latter value could be discovered and inserted in the mathematical expression, it would give a reliable scientific value for the unknown age of the earth.

The brilliant concept by Brown arose intuitively from his previous mass spectrometric studies of isotopic abundances of traces of cobalt in meteorites and their relationship to primordial compositions of stars. He was led to develop, from knowledge of chemical compositions of stone and iron meteorites reported by previous investigators, the idea that the primordial isotopic composition of lead in the solar system could be measured in lead obtained from iron meteorites. Prior to the war, Ida and Walter Noddak had worked out general knowledge of macro and micro chemical abundances in meteorites as the foundation for understanding the chemical composition of planets in the solar system. Brown became familiar with their work through his cobalt isotope studies. The difficulty with Brown's idea was that the Noddaks' work showed the existence of only micro concentrations of lead in iron meteorites.

Brown also developed concepts for pioneering mass spectrometric methods that could determine geological lead/uranium ages of common igneous rocks by grinding some kilograms of them into powders and isolating for analyses a few milligrams of the uranium-rich microminerals, such as zircon, that the rocks contained. Here again, the difficulty lay in the existence of only micro quantities of uranium and lead in these few milligrams of zircons. Brown persuaded his graduate student, George Tilton, to develop chemical and mass spectrometric isotope dilution techniques for measuring micro amounts of uranium in such minerals, and he persuaded Pat to develop chemical and mass spectrometric techniques for measuring micro amounts of both concentrations and isotopic compositions of lead in them. Combinations of these uranium and lead data would yield ages of the rocks. At the end of the war there were available only a handful of radiometric ages of common rocks for the entire earth, and these had been determined in macro minerals obtained from rare uranium ore deposits whose times of formations were associated with formations of those rocks.

Tilton and Patterson worked together with mineral separates of zircons obtained from an igneous rock magmatically related to and of the same age as a uranium ore deposit that had been previously and accurately dated by Nier. Macro chemical and mass spectrometric techniques were applied to gram-sized minerals by Nier for this purpose. The zircon microminerals were given to Brown by Esper Larson, a Harvard professor of geology who had obtained them as part of his effort to expand geochronometric knowledge. Tilton and Pat had neither knowledge of such mineral separations nor geology. They belonged to a third generation of geochemists, represented in the first by Aston who developed the mass spectrometer, followed in the second by Nier who worked out radioactive decay parameters and isotopic abundances, and then in the third by Brown, their mentor, who pioneered the new concepts of radiometric dating. It was also Brown, together with Professor Mark Inghram, co-mentor with Brown of Tilton's and Pat's graduate work, who designed and built in a Physics Department building the new type of mass spectrometer needed for these studies. Working with milligrams of minerals separated from kilograms of rocks, Tilton and Pat were able to devise techniques dealing with micrograms of uranium and lead that provided the first accurate radiometric age of a ubiquitous type of igneous rock. Today, fourth generation geochemists work with nanogram and picogram quantities of lead and uranium in single microcrystals.

5. UNIVERSAL LEAD CONTAMINATION

Tilton found hc could not work with microgram quantities of uranium in laboratories where gram quantities of uranium had been studied previously, because of contamination. He solved some of his contamination problems by moving to a laboratory virgin in this respect that was in a Geology Department building away from the Chemistry Department building. Here, he refined his techniques to arrive at a successful method for measuring concentrations of micro amounts of uranium in microminerals of common rocks by mass spectrometric isotope dilution. However, Pat's lab in a Chemistry Department building was ubiquitously contaminated by industrial lead originating from numerous sources common in all laboratories that had never been previously recognized. Many of his techniques for controlling lead contamination were developed as a consequence of dependent relationships with Tilton's uranium measurements.

Patterson and Tilton's U/Pb dating studies of rock microminerals, together with Nier's previous macro studies of uraninite in an associated pegmatite, provided an accurate age for the rock. Combinations of Tilton's measured uranium concentrations in the same microminerals Pat analyzed, and the known age of the rock, dictated specific isotopic compositions and concentrations of lead in those minerals. When Pat made mistakes in lead contamination control in his mass spectrometric measurements of lead, they showed up as erroneous deviations in his measured isotopic compositions as compared to the required calculated compositions. Pat had to keep refining his contamination controls until those incorrect deviations disappeared. He proceeded to meticulously identify each and every contamination source by tracking isotopic changes in a lead isotope spike as it progressed through a procedure. Extraneous industrial lead, originating from reagents, containers, air, handling, and collection, had to be separately identified and reduced to bring total contributions down to known, acceptable levels.

George Tilton and Pat were finally able to report in their PhD theses and scientific journals new techniques that could be used by others for dating ages of common igneous rocks. With knowledge of their new techniques widely available, the field of geochronometry became enormously expanded, which in turn greatly accelerated the development of new and better geological knowledge.

Simultaneously, Pat's discoveries of the ubiquitous and extensive nature of industrial lead contamination, and more important, how it should be correctly dealt with, allowed him to obtain the true concentrations of lead in various substances. This would later become a decisive factor in disclosures of widespread errors in reports of lead occurrences, as well as in concepts regarding natural lead levels and modern lead contamination by man.

6. Dating the Age of the Earth

Brown had encouraged Pat as a graduate student to develop a method for measuring the isotopic composition of lead in iron meteorites which could be used to determine the earth's age, exclaiming "It will be duck soup and you will be famous!" Both statements turned out to be in serious error. Pat discovered that Brown's idea, based on the Noddacks' reported lead concentrations in iron meteorites, involved mistakes in the Noddacks' lead measurements. Because of industrial lead contamination unrecognized by them, they obtained values for trace lead concentrations in iron meteorites that were some fifty times higher than true values. Much of Pat's contamination control techniques had been worked out earlier in rock dating studies that involved amounts of lead that were thousands of times smaller than those studied previously in chemical laboratories. However, he then had to extend the scope of those contamination controls and develop them further in the isolation of lead from iron meteorites, because both sample sizes and amounts of lead contamination originating from chemical processing were magnified by about 50-fold. Pat had to laboriously identify each of the numerous new sources of contamination, then determine their quantitative amounts, and then successfully eliminate each contamination separately in order to finally isolate uncontaminated micro amounts of meteoritic lead that was required for mass spectrometric analysis. This required many years of work dedicated to integrity and unyielding to compromise which was to become a pillar of the Patterson geochemical heritage.

Pat submitted a proposal to the US Atomic Energy Commission requesting funds that would enable him to carry out the dating of the earth as a postdoc at Chicago. The AEC had been funding the earlier part of his work previously through Brown while he was Brown's graduate student. He wrote in the proposal how he could finally obtain a scientifically meaningful value for the age of the earth by measuring the isotopic composition of lead in iron meteorites. However, the proposal was rejected by the AEC, for that earlier work was funded in a context of developing knowledge of future sources of uranium fuel for power. Brown, with political acumen, resubmitted the proposal in his name, explaining how Pat's work would be potentially useful to the AEC because his new analytical techniques could be used for future studies of labile uranium in igneous rocks. Brown pointed out to them that this was extremely important because amounts of uranium leached by weak acids from one ton of pulverized granite (Tilton had already obtained preliminary data on one granite) could provide the energy for atomic power plants equal to the amount of energy supplied by ten tons of coal in conventional power plants. Brown's proposal was funded, which allowed Pat to work as a postdoc during 1951-52 at the Institute for Nuclear Studies at the University of Chicago, and then continue later in the Division of Geology at the California Institute of Technology.

Pat proceeded to chemically separate lead from iron sulfide phases contained within a brother of the Grinnell fragment of the Canyon Diablo iron meteorite that had been obtained by Brown. This separation occurred in a new ultra-clean lab he built at Caltech, and the isolated lead samples were flown by Pat to Chicago, where he analyzed them in a new mass spectrometer developed by Inghram. After seven years of identifying and eliminating various sources of industrial lead contamination in the laboratory (work that began with Brown, Tilton and Inghram at the University of Chicago and culminated at Caltech), Pat finally measured the isotopic composition of primordial lead in iron meteorites. This enabled him to derive the first conclusive age for the earth and solar system. This measurement established at this time the first definitive boundaries for the geochronology of the solar system. Equally important, these techniques made possible future analyses of lead at low levels encountered in environmental samples of air, waters and biota.

Only a handful of geochemists among the world's total population of geologists understood the meaning of Pat's measurement of the age of the earth when he reported it in 1953. During the next several years, other methods were developed for determining the age of meteorites. Patterson obtained further lead isotopic data for stone meteorites, which allowed a better view of the age of the earth and solar system. However, because nuclear geochemistry was a new post-war entry into the ancient science of geology, and because of its foreign character, many of its findings were not immediately recognized nor understood by most contemporary geologists. It took a decade and a half before Pat's pioneering value for the earth's age began to appear in newly published geologic texts, more or less as an abstract parameter. Even then, mention of the manner of its determination, sometimes given as brief asides in those texts, were usually erroneous and misleading. Patterson's name was never generally recognized in association with that determination, although the age value has changed little through subsequent decades. Even one of Pat's closest colleagues whose work among others in the field verified the first measurement during later years of effort, exclaimed to him, "Pat, you didn't measure the age of the earth! We did! God simply handed those lead numbers to you!"

7. Lead Sagas from the Ocean

Pat's scientific career was nurtured in the geology department at Caltech, where he learned geology and its relationships with lead isotopes in the earth's crust from Leon Silver, Bruce Doe and Rama Murthy. These colleagues guided Pat's research interests into developing understandings of origins of the earth's crust from its mantle using lead isotopes as tracers exhibiting different characteristic temporal changes in lead isotopic compositions. These in turn could be related to different Pb/U ratios characteristic of mantle components generating rocks of acidic, intermediate and basic compositions.

In connection with these studies, Pat and his colleague T. J. Chow studied chronologic changes in isotopic compositions of lead in continental rocks of different ages and chemical character. This was done by examining their isotopic records stored in drill cores of ocean sediments taken over large areas. They reconstructed time sequences of aerial patterns of lead isotopic compositions over the ocean floors that were contributed by rocks of different ages from different continental regions at different times. This work was supported by the American Petroleum Institute whose company members were interested in using temporal changes in lead isotopic compositions recorded in sediments to help identify strata and sources of petroleum. Pat and Chow found that two types of lead existed in marine sediments which could be examined separately: (1) that which was carried in clay particles from rivers to ocean bottom sediments; and (2) that which was soluble river water lead initially dissolved in sea water before being incorporated in biologic particles. Although such particles formed in upper ocean waters, they were rapidly deposited in sediments where lead was remobilized into authigenic minerals. They obtained ages for strata in those sediment cores from Ed Goldberg who had derived them from use of his newly developed ^{230}Th dating method. Combined with their observed concentrations of deposited authigenic lead, they were able to quantitatively measure what the flux of this authigenic lead from rivers through the ocean to its sediments had been during past eras. This latter fraction was part of the authigenic iron precipitates and could be simply leached using moderately strong acids.

Then Pat and Mitsunobu Tatsumoto started to interpret these patterns by working out theoretical models relating transfers of uranium and lead from the earth's mantle to its crust with histories of changes in isotopic compositions of lead in various types of crustal rocks. At the same time they began to measure concentration profiles of lead with depth in ocean waters, and inadvertently made an important discovery - that upper ocean water layers contained much more lead than deeper layers. Pat calculated present day world-wide inputs of dissolved lead to the oceans, estimated from lead concentrations in river waters reported by other investigators. Reported lead concentrations were 50 to 100-fold in excess of amounts of lead that had flowed from rivers through the oceans to their sediments in the ancient past, as determined from authigenic lead fluxes in sediments Pat and colleagues had measured.

Pat suspected that the river water measurements were erroneously excessive due to unrecognized contamination (he and his colleagues later showed this was so), so he shifted to consideration of leaded gasoline production data to calculate how much industrial lead could be transferred from leaded gasoline exhausts to widespread atmospheric aerosols which then become incorporated in rain that fell on the oceans. Pat and Tatsumoto published a paper showing that excessive atmospheric inputs of lead to the oceans were highly probable, enormously exceeded ancient natural fluxes of soluble authigenic lead through the oceans, and could account for the excess lead observed in upper oceanic water layers. This discovery was to profoundly and permanently alter the direction of all Patterson's subsequent scientific research (Figure 2).

Figure 2. Pat Patterson at mid-career at his Caltech laboratory. In the background is the still which produced the purest of water necessary for the accurate blanking of trace element clean procedures which he pioneered.

8. At Odds with the Oil and Lead Industries

At this time a major portion of Pat's research had been supported by the American Petroleum Institute, and his paper with Tatsumoto clearly implicated the Ethyl Corporation and its oil refinery customers as major contributors to serious lead contamination of the atmosphere and biosphere. This was to put Pat on a collision course with the oil and lead industries.

Economic factors requiring the use of lead alkyls in gasoline forced the oil and lead industries to try to counteract and suppress the generation and dissemination of global lead contamination information. The oil industry tried to bribe Pat, but in meetings with him, he responded negatively by lecturing them with conclusive illustrations. Pat led discussions of newly conceived theoretical models in which he proposed how investigators could use such global lead contamination in future experimental research to develop scientific knowledge. He tried to convince the industry on the manner and extent to which leaded gasoline exhausts were probably contaminating terrestrial and marine food chains and humans to a degree far above preindustrial natural levels. Contamination was on a world scale and lead-poisoned environments existed everywhere.

The petroleum industry response to these lectures was to terminate his API funding. They also encouraged some members of the Board of Trustees of Caltech who were officials of oil companies to try to have Patterson fired from Caltech, as at the time he was only a postdoc without tenure.

9. The Forging of an Environmental Phoenix

Angered by such industrial tactics which put his traditional professional career in doubt, Patterson's research began a rebirth into the environmental aspects of lead. He was confidently aware of both the probable (but as yet unrecognized) excessive lead contamination of the environment as shown by excess lead in the oceans, as well as it being hidden by a widespread and excessive (but unrecognized) lead contamination in laboratories. Pat's initial focus of interest was on generating new ideas of how to develop scientific knowledge about environmental lead.

Harriet Hardy, a toxicologist on the faculty at MIT, encouraged his efforts in this direction during his year at MIT as a visiting faculty member in the Geology Department in 1963. He read in materials obtained from the Harvard Medical Library that most lead and barium in people was found to reside in their skeletons. Of these two very toxic trace metals, only lead was extensively used industrially and introduced into the environment. He also read of findings that ratios of radioactive strontium from bomb fallout to calcium in cattle fodder were much greater than in cow's milk, due to less efficient uptake of strontium compared to calcium along nutrient pathways. Assuming that both lead and barium followed metabolic pathways similar to calcium, Pat developed the concept that metabolic calcium should be biopurified of metabolic barium and lead to greater and greater degrees by precursor consumers at consecutive trophic levels in food chains leading to humans. He made theoretical predictions of what should be the natural Pb/Ca ratios in soil, foods, and humans by reducing them by the same factors Ba/Ca ratios were observed to be reduced in those materials. According to his new concept calcium in humans had indeed been biopurified more than a hundred-fold of barium as it traveled along food chains in precursor plants and herbivores from rocks to them. However, observations of Pb/Ca ratios in contemporary humans showed that their bodies contained, astoundingly, the same ratio as did the rocks which served as primary nutrient sources of calcium in food chains and humans. This meant that natural processes which biopurified calcium of lead had been overridden by industrial lead contamination. Moreover, because the lead had been added through unnatural means to such an extent, humans had poisoned themselves with a hundred-fold greater amount of lead than they should have consumed naturally.

Katherine Boucot, editor of a prominent scientific journal used extensively by industrial toxicologists to report their findings, knew of Pat's new interests through her son who was a professor of geology on the Caltech faculty. She asked him to write an expert's comment for her periodical and he responded by asking her to publish his theory, which she did. This greatly exacerbated public arguments then developing between lead industries and groups who wished to control the use of lead. Lead poisoning among people at large was becoming a political issue at this time through efforts by activists to eliminate use of lead in paints and gasoline. The publication of Pat's biopurification theory from the same citadel journal which published industrial toxicology's dogma was helpful to the anti-lead activists as academic scientific ammunition in their behalf. As an earth scientist, Pat tried to remain aloof and did not participate actively in further debates in this matter, although he did testify on behalf of the anti-lead interests at various national and state legislative hearings.

Eventually, efforts by activist opponents of public uses of industrial lead gave rise to enactment of legislation requiring reduction and eventual elimination of lead in gasoline and household paints. However, authorities attending an international meeting of industrial toxicologists held in Europe at that time addressed the controversy. They declared Patterson's view to be a fanciful academic myth that contradicted what they viewed to be the realities within toxicologically meaningful environments in which humans actually lived. They stated that, contrary to his assertion, typical levels of lead in people are perfectly normal, and have been observed not to be associated with poisonous effects. The lead industries exerted political pressure on bureaucrats in the former US Public Health Service. Through academic industrial toxicologists whose work was funded by the lead and oil industries, any requests by Patterson to the PHS for research support were denied.

10. READING THE ICE CUBES

Pat returned to Caltech in 1964 to concentrate on his theoretical, field and laboratory research efforts to develop scientific knowledge of natural and industrially perturbed cycles of lead in the atmosphere, biosphere and oceans. A critical goal in this matter was to obtain knowledge of the historical record of changes of lead concentrations in the earth's atmosphere. Goldberg had at that time developed a ^{210}Pb chronometric method for use in Swiss Alp glaciers that allowed assignment of specific ages to ice strata. This led Goldberg to point out to Pat that the best record lay in snow and ice accumulating at latitudes and altitudes where it did not melt and thus preserved thousands of years-old relics of past atmospheric chemistries. Pat seized upon this idea and developed, with his colleague Masayo Murozumi, methods for ultra-clean sampling, transporting, and then analyzing samples of glacial ice for lead. This was an arduous and tedious task, since lead concentrations in field samples were less than those reported by most investigators in distilled water. It involved huge sample sizes, containers, and chemical extraction apparati at unprecedented levels of geochemical cleanliness. The result was the ultra-clean digging of deep shafts in Polar ice caps, and extracting large ice blocks from the walls of these shafts. The ice blocks were removed by Pat and his colleagues wearing acid-cleaned plastic suits and gloves and using acid-cleaned equipment brought to the site. Wrapped in acid-cleaned plastic, the ice blocks were then placed in acid-cleaned plastic tanks at the site for transport and subsequent processing in special ways in the Caltech clean laboratory (Figure 3).

The US Army Corps of Engineers and the US Air Corps supported the logistics of flying these bulky ice samples in and out of Greenland. The US Navy Seabees supported the logistics of flying the samples out of Antarctica. In the Antarctic, there was a traditional Ice Bowl football game on New Year's Day between the scientists and the Seabees. Pat's New Zealand crew was allowed to join the scientists' team for the game, and after hasty instruction and practice on the Ice, managed to win for the first time. However, Captain Bligh (Pat) Patterson permitted his men to participate in the game only after they had finished digging the deep shaft, to ensure that possible football injuries did not interfere with the real work.

The US Atomic Energy Commission supported the analytical work associated with Patterson's collection of polar ice. The result was the now classic profile of changes in lead concentration in Greenland ice during the past three centuries obtained by Murozumi, Chow, and Patterson and published in 1969. These data clearly showed increases of lead in the atmosphere of the Northern Hemisphere after the Industrial Revolution together with global dispersion of atmospheric lead to remote regions of the hemisphere. This was followed by enormous increases of atmospheric lead concentrations beginning in the third and fourth decades of this century after the use of leaded gasoline began.

The meaning of this work was not accepted by the scientific community for years because the values were many orders of magnitude less than established values, and others could not verify the results when they tried to do so. But these findings were authenticated a decade later by Pat and Amy Ng, using more sophisticated and reliable techniques that could determine lead in samples of ice that were 100-fold smaller. This allowed them to study Greenland ice cores collected earlier by other investigators. They published in 1981

Figure 3. Pat Patterson in Greenland during the extraction of clean snow and ice samples in which he first chronicled the global extent and history of lead pollution.

that exterior portions of these ice cores that had been studied by other investigators were contaminated with industrial lead added during drilling and handling. These exterior samples had levels of lead that were 100,000-fold above those in the non-contaminated interior samples. This disparity in part accounted for the inability of others to verify the lower Caltech findings. The lower findings became generally accepted internationally several years later among colleagues at various universities in the US and other countries, most of whom journeyed to Pat's clean lab at Caltech for training. In addition, several international scientists interacted with Pat in their investigations of geochemical cycles of trace metals and learned the art of clean lab research and mass spectrometer analysis.

Claude Boutron and Pat used similar techniques to study ice samples from Antarctica to show industrial lead had indeed contaminated the atmosphere even in the remote Southern Hemisphere, but in lesser amounts in accordance with the fact that most industrial lead had been introduced to the earth's atmosphere in the Northern Hemisphere. Boutron later used such techniques to extend knowledge of temporal changes of atmospheric lead in Greenland ice during both the last century and during earlier centuries of Greek and Roman lead smelting activities. These data showed a decline of atmospheric lead in recent decades due to restrictions and elimination of use of leaded gasoline. They also showed peaks of elevated atmospheric lead concentrations corresponding to excessive lead smelting operations two millennia earlier during Greek and Roman times. Indeed, Pat had predicted this global lead insult during these ancient times from calculations of Greek and Roman lead and silver production, and had published this information two decades earlier.

11. Mountain Vistas in Lead Colors

Patterson and his colleagues Todd Hinkley, Yoshi Hirao, Cliff Davidson, Robert Elias, and Yigal Erel studied natural flows of rock lead through soils, plant and animal food chains, air and waters in a remote, high altitude mountain canyon located in a pristine region of the Sierra Nevada. From a decade of measurements of lead concentrations and isotopic compositions in these substances and in soil and pond sediment layers, they made startling new discoveries of the extents to which industrial lead had entered the canyon from the atmosphere during the past century and had contaminated it. They found that most of the lead in litter, sedge, and the top one centimeter of soil was of industrial origin, as was the lead in voles (herbivores) and martens (carnivores). Furthermore, industrial lead made up about one-fifth of the lead in ground water and nine-tenths of the lead in snow melt flowing out of the canyon. The annual input of industrial lead to soil humus of the canyon was 10 times greater than the natural lead introduced by rock weathering. They found, in experimental verification of Pat's earlier biopurification theory, that calcium was biopurified of barium 1000-fold in going from rock to carnivore, actually a 10-fold greater factor than his theory predicted. Also, calcium was biopurified of lead by this same factor, but only after measured additions of industrial lead at various points along the pathway, which summed to 99% of the lead in carnivores, had been subtracted.

12. The Ocean Gives up its Lead

During Patterson's studies of lead cycles in the oceans, Dorothy Settle became his long-standing co-worker who helped organize and carry out research planning, field operations, and laboratory investigations with graduate students, postdocs, and visitors from domestic and foreign universities. By and large, these investigations were carried out within government sponsored, domestic and international inter-university NSF research projects, such as IDOE, SEAREX, and AEROCE and included P. Buat-Menard, T. Chow, T. Church, R. Duce, W. Fitzgerald, A. Véron, and C. Wong. These activities involved cooperative field sampling operations followed by inter-laboratory exchanges of results and interpretations.

Early in this period Pat came in contact with some engineers engaged in a US government project to drill through Antarctic ice to allow investigations of ancient waters in an ocean basin that had been sealed by ice and isolated from industrial pollution. During interactions with them, Pat developed the idea that they should pause and lift their drill out of the hole just before they broke through its bottom into pristine ocean waters below. Through this contaminated drill section they could lower a new type of water sampler that he had designed, and it could be allowed to break through and be lowered a few tens of meters further. Upon a signal, it could open up a protective shell that encased its uncontaminated interior and allowed it to sample virgin waters beneath the ice. Although investigators in that project scorned and rejected this idea, B. Schaule used it to successfully develop a new sea water sampler which he and Pat used to carry out studies of lead in sea waters. The sample chamber was encased in a protective shell that did not open until it had descended below contaminated waters surrounding an oceanographic research vessel. Tatsumoto and Pat's earlier lead concentration profiles were found by this new device to be erroneously high because their sampler had been dirtied by lead contaminated waters surrounding the research vessel. Sea waters collected in this new manner were found to contain 100-fold lower lead concentrations (less than in distilled waters found in most laboratories) from the same ocean areas Pat, Tatsumoto and Chow had studied previously. Surprisingly, shapes of depth profiles of lead from those various oceanic locations remained the same as before, showing very large excesses in upper water layers less than one kilometer deep with new, extremely low and relatively uniform concentrations in deeper waters . These lead excesses were greater at similar latitudes in the North Atlantic than in the North Pacific, and they decreased as one moved southward from the central Pacific. This geographic pattern was in accordance with industrial production and use data indicating greatest emissions of atmospheric lead moving eastward from North America. This compares with lower emissions of atmospheric lead moving eastward from Asia, with the smallest emissions across the equator in the Southern Hemisphere.

In discussions with Schaule, Pat reminded him of the historical background of their findings by holding a graph of their new picogram lead per gram sea water results and pointing to the edge of the high concentration bulge in the profile curve image. Pat exclaimed "OK, on the scale of this plot of our new data, Tatsumoto's and my earlier values are indeed located out the door of this room and part way down the hall. But remember and be prepared. In earlier years, on this same plot scale, older values for lead in sea water found by all previous investigators are located out of this building and then way across to the other side of the campus! Those guys are the type who will be evaluating our findings, and few, if any, will believe them worthy, because earlier they found it difficult to accept even the down the hall data!"

However, fellow oceanographers within their inter-university research group did value the findings and used similar versions of the sea water sampler plus ultra-clean sampling and analytical techniques developed by Pat's CIT group to benefit their field and laboratory studies of other trace metals in the oceans. Perhaps most notable was the late J. Martin and his colleagues who were enabled, through resulting improvements, to make the first reliable measurements of iron in sea waters. They discovered that inputs of atmospheric dusts to open oceans probably supply major portions of nutritive iron in those waters and so might regulate magnitudes of primary production in otherwise nutrient replete oceans. This, in turn, could have feed-back relationships with global atmospheric chemistries including CO_2 and, thus, climatic changes during glacial-interglacial eras.

M. Burnett, R. Flegal and Pat found that calcium in sea water, after incorporation into phytoplankton, became biopurified of lead as it was passed along successive trophic levels of marine food chains, just as was the case in terrestrial food chains. As part of this work concerning consumers near the ends of such chains, Pat and Settle found that tuna, because they lived as carnivores in marine environments that were less contaminated with lead than terrestrial ecosystems, possessed muscle that contained less lead than any other type of animal muscle in the world. Pat and Settle then determined lead in a US NBS tuna muscle interlaboratory standard and also sent a sample of their tuna muscle to the US FDA analytical laboratories for lead measurements by them. Subsequent discussions ensued between Pat and the US NBS regarding their certified lead concentrations in their extremely lead contaminated tuna muscle standard. Similar discussions occurred between Pat and the US FDA regarding 10,000-fold positive errors in their analyses for lead in freshly butchered tuna muscle before it was placed in lead soldered cans. This became an important factor in causing the US FDA to later work out an agreement with manufacturers of all canned foods to switch from lead soldered to electrically welded lead-free food cans.

Amy Ng, Settle and Pat showed that a greater fraction of lead inputs via rivers from cities to ocean coastal waters originated from storm washings of gasoline lead exhaust aerosols deposited on roofs, paved surfaces and trees, and from manufacturing outputs in sewage outfalls. They found that increases of such lead during the past half century were recorded in chemically labile lead of coastal sediments, showing recent maxima at different locations ranging from 2-fold to 50-fold above pre-1900 concentrations. They found that coastal waters and marine organisms living in those waters near urban regions were extremely polluted with lead (compared to organisms living in mid-ocean waters). Also, much of the lead contained in these waters was removed by biological activity and sedimentation to coastal shelf deposits before currents could transport it out to open ocean regions.

This meant that excesses of lead in mid-ocean surface waters came from the atmosphere, and Buat-Menard, Church, Erel, Flegal, Maring, Rosman, Settle, Véron and Pat worked out the various ways and their magnitudes by which atmospheric lead entered different ocean regions. They collected aerosol, dry deposition, rain and sea water samples from towers on remote islands and from ships at sea. They found that industrial lead-rich carbonaceous-oxide aerosols are scavenged from air by rain and added to oceans less efficiently than are lead-containing soil dusts. A large fraction of this added lead was found, through isotopic tracing, to linger in thin, oily films on the ocean's surface, where bursting bubbles from breaking waves formed it into second-generation lead-enriched seasalt aerosols that were returned to the ocean through efficient scavenging by rain.

They also analyzed lead and sulfur in volcanic gases from various active craters on different continents, then used their observed Pb/S ratios to measure amounts of volcanic lead entering the oceans via rain. From co-existing sea salt and non-sea salt sulfate concentrations in rain, the flux of volcanic sulfur to the atmosphere had been previously determined by other investigators, but now could be used to determine global fluxes of natural volcanic lead to the atmosphere. They measured and compared amounts of industrial lead entering the oceans in carbonaceous-oxide aerosols, in soil dusts, in plant leaf aerosols, and in recycled sea spray salts, and compared these to amounts of natural lead entering the oceans in soil dusts and volcanic aerosols. They measured ways in which proportions of these different inputs varied at different oceanic locations. They established new and more meaningful values for relative enrichments of sea salts, soil dusts and industrial lead aerosols incorporated from air into rain. Along with B. Glover and K. Rosman, they analyzed samples of air, rain, fog, dry deposition, and seawater to establish values for the fluxes of natural and industrial lead to the North Atlantic and to the North and South Pacific Oceans. In this way, they were able to identify relationships between variations in industrial emissions from land, in air mass flow, and in magnitudes of input fluxes. They also delineated mixing pathways of surface sea waters by use of lead isotope tracers.

13. The Truth Is In Our Bones

In research funded by NIEHS, J. Ericson, H. Shirahata, M. Manea-Krichten and Pat studied occurrences of lead in bones of ancient New World Indians. They found that original biological levels established during the lives of these ancients had become overridden and obscured by huge amounts of soil moisture lead added during many centuries of burial. They compared occurrences of lead among different types of bone with occurrences of barium and calcium, and established relationships between distribution patterns of these metal concentrations and sizes of apatite crystals. Then they and G. Miller, Settle and Erel studied occurrences of these three metals in fresh bones from modern cadavers to establish these same relationships on a clear biological basis in bones that had not been exposed to prolonged perturbations from soil. They worked out, from comparisons of these two sets of occurrence patterns, what natural levels of lead, originating from rock-lead in food, had been in preindustrial humans, and found those levels were about 1/1000th of typical levels in people living in industrialized regions of the world today.

14. Toward New Lead-Free Biochemical Knowledge

Pat and Settle had also measured the extent of industrial lead contamination in contemporary biochemical laboratories by collecting samples of various tissues and nutrient media being used in such environments. After analyzing them for lead, and comparing their findings with natural lead levels in analogs of these materials, they inferred corrections made from their extensive measurements of lead contamination effects in food chains of remote ecosystems. They found lead contamination factors ranging from 10-fold to 10,000-fold above inferred natural levels that existed in materials being studied in biochemical laboratories. This had profound implications regarding all existing biochemical knowledge of lead, for it meant that this knowledge had not been developed from studies of uncontaminated biological systems. The implications were that large excesses of lead levels in typically contaminated biological systems had been perturbing natural biochemical processes in various unknown ways, which obscured knowledge of natural biochemical processes. Scientists within the biochemical community were completely unaware of the ultimate meaning endowed by the existence of widespread industrial lead pollution. They were only vaguely aware of it in the first place, and, as an aside, believed it affected things outside their laboratories but not their own experiments.

So Pat and Settle spent several years preparing a proposal to NIEHS to carry out cooperative studies with biochemists already at Caltech. This group would develop methods for growing colonies of rats that contained both very low natural levels of lead and also colonies that contained far lower, unnaturally ultra-low levels of lead. They would develop methods for obtaining preliminary clues to principal characteristics of new mechanisms in lead biodynamics at these very low natural and ultra-low unnatural levels. In principle, this would provide proper conditions and controls for investigations of natural biochemistries which their fellow biochemists would then be able to pursue. This proposal involved extremely meticulous, time-consuming and expensive developments of new techniques that could be used for studying lead in biological systems at levels far lower than any ever studied before. The goal was to develop new techniques for controlling lead contamination at such levels. This would enable them to correctly determine lead in biological materials at those levels and to devise rat foods containing those levels of lead. Also, it would illuminate critical mechanisms in a few new previously unknown natural biodynamic pathways at low lead levels that biochemists had never thought of before, and to assist training biochemists in this field educated in new techniques for preparing the animals and studying their biochemical systems at appropriately controlled lead levels. In this manner development could begin of new scientific knowledge of natural biochemistries unperturbed by industrial lead. At the same time new studies of systems, ultra-purified of toxic trace metals far below natural levels, might point out clues regarding differences between living biochemical systems in cells and similar but non-living chemical systems in test tubes. The assumption is that hierarchies of interrelated biochemical signals and biodynamic processes had previously functioned properly in a context of natural toxic metal concentrations within cells before global industrial metal pollution. Now these hierarchies might become measurably confused and scrambled in chemical silence from those toxic metals produced in ultra-purified laboratory animals, calling experimental attention to their existence and functions that had been previously unrecognized. During the next several years the proposal was rejected, revised and resubmitted, and rejected again, on the basis of being awarded the lowest of merit ranking scores, first by veterinarians and then by molecular biochemists.

These reviewers felt the excessive time and expense of the proposed technique development work was unnecessary and unjustified. Further, they viewed it as of secondary importance in the proposed matter since such things could be readily done quickly and inexpensively through simple instructions for modification of procedures to technicians. More important, proposed studies of the mechanisms mentioned in the proposal were too elementary because they were already being studied by more qualified investigators at much more sophisticated levels. The reviewers believed that instruments could and did provide answers to scientific questions. However they were unaware that this attitude was the exact reason why nearly all investigators were unable to measure lead concentrations correctly. They failed to understand that their knowledge of "mechanisms in biodynamic processes" applied only to a lead-perturbed context. They mistakenly believed that lead contaminants in their laboratories were having no effect on these processes. Furthermore, these lead effects were never consciously questioned at each and every point in interpreting their findings. They failed to understand that an unrecognized non-lead perturbed natural experimental context imposed quite different, alternate meanings to biochemical knowledge.

15. How and Why the Brain is Wired

So Pat, in the end, switched all of his inquiries to study those factors that caused humans to poison the earth's biosphere with industrial lead. During previous decades he had studied archaeological findings concerning developments in metallurgical technologies. These were initiated by the beginning of agriculture at the onset of the Holocene 12,000 years ago, that gave rise to mining, smelting and use of lead and silver during the Greek, Roman, Industrial Revolution and recent eras. He constructed tables of world lead productions for those three earlier periods, generated by a mixture of ideas that used data from a number of different sources. These included published data concerning quantities of ancient Greek and Roman slags left from their operations that had been reworked in modern times to recover residues of silver remaining in them, as well as relative values of lead and silver during ancient times. These also included published data concerning amounts of silver shipped from the New World to Europe by Spaniards during the 16th and 17th centuries, loss rates of silver coins in modern times, and published compositions of lead-silver ores. Recent studies of variations of lead concentrations in annual layers of Greenland ice and Scandinavian pond sediments showed peaks and valleys, resulting from variations of additions of lead smelter fume to the atmosphere during the past two thousand years. The changes in magnitudes of concentrations corresponded with changes in magnitudes of world lead production given in Pat's tables published two decades earlier.

Pat worked during these recent years with John Rowe and Chris Donnan to chemically analyze culturally associated metal artifacts recovered from sites in Meso and South America in order to delineate temporal sequences in uses of ore types and smelting techniques. Pat found, using these data in combination with archaeological data provided by other investigators, that cultures in these regions had achieved three major stages in advances of metallurgical technologies over four thousand years. Moreover, these advances were identical with those achieved by cultures in Eastern Mediterrania during a similar four thousand year period that began five thousand years earlier.

Pat also found that the beginnings of science had been born in an environment of business enterprise that had been fostered in Greek cultures by the invention and use of silver coins made from outputs from huge local lead-silver ore deposits. This depended on developments in metallurgical technologies within serf-agrarian empires in Eastern Mediterrania over thousands of years. He showed how the growth of modern science, after discovery of the New World, depended upon silver produced and obtained from lead mines in South America. Infused with a huge burst of new silver coin, Europe was stimulated toward manufacturing enterprises which ignited the Industrial Revolution. He pointed out how this gave rise to developments in various kinds of engineering technologies and financial supports which nourished, through feed-back relationships, the establishment of science as the effective social institution it is today.

Consideration of this historical evidence led Pat to formulate some important new concepts regarding the manner in which the human brain functioned as it developed different kinds of knowledge during the past 12,000 years. The first of these concepts is the huge potential for complex, higher order functions whereby neuronal circuitries were unintentionally installed by biological evolution. These remained unused in tribal environments until brain population densities increased to critical levels within urban environmental contexts. Although these urban centers were created by greater and more stable food supplies, they were unnatural with respect to natural biologically evolved contexts for all previous ancestral hominids. Development of complex hierarchies within social institutions, coupled with associated growth in complexities of communication, engendered rapid development in complexities of cognition, language and knowledge during the following 10,000 years in regions of high population densities.

Pat's second concept postulates that sophisticated abstract ratiocinations carried out in the *Homo sapiens sapiens* brain are of two different kinds, utilitarian and non-utilitarian. These involve biochemical functions within two geometrically separate and different neuronal circuitries within that brain. This originates from his observation that identical developments in metallurgical technologies were carried out by two different, regionally isolated populations, while during this same time sharp contrasts exist in cosmogonies, music and writing developed by the two cultures.

Pat's third concept developed from considerations of relationships among histories of technologies, religion, governments, and arts and sciences. This is that utilitarian ratiocination has, for the most part, determined the character of social institutions, technologies, cognitive structures of languages, and knowledge as they developed through feed-back interrelationships within successive cultures. He postulated that non-utilitarian ratiocination has played a subservient role in this operation.

Pat suggested how brain-scan investigations could be made that might show differences between neuronal wiring used for non-utilitarian abstract ratiocination and neuronal wiring used for utilitarian ratiocination. Up until his death in December 1995, he devoted all his efforts to the preparation of a book outlining to young scientists the kinds of research they might carry out to delineate various kinds of environmental and emotional factors involved in cultural developments during the past 10,000 years. The purpose of this work would be to determine relative strengths of utilitarian and non-utilitarian thinking. In this way he hoped that a small number of scientists working in fields of anthropology, archaeology, history, philosophy, religion and biochemistry might form a new type of scientific knowledge that would disclose the difference between how humans think and act today, and how they should think and act in ways that would show the worth of humankind transformed into reality. An abbreviated version of this last unfinished work by Pat entitled "Scientific Meaning of Meanings; Quests for Discoveries Concerning our Cultural Ills" was published in a special volume of *Environmental Research* edited by A. R. Flegal.

16. A Prologue on Scientific Wisdom and Integrity

In early years of his work Patterson gathered some bird droppings of various kinds and shapes and glued them to the outsides of the magnets on his mass spectrometer in a graceful, more or less realistic, artistic display. According to traditional ritual, Pat's colleagues entered the instrument room containing the spectrometer, carrying their precious lead samples for analysis they had isolated in the chemistry lab with exhaustive effort to prevent contamination. They then inserted the samples into the instrument, using ultra-clean procedures and paused while the machine made itself ready for analyzing the sample. Repeatedly, they were confronted with this most foul visual desecration of their sacred samples. The first time each of his colleagues saw this spectacle, they were horrified, and rushed to Pat to have it removed. But Pat explained it was there, and must remain there (it did not affect the samples), to remind them of the ultimate factor governing their efforts to acquire scientific knowledge. His message was that the instrument room contained analytical machines that were only slaves of humans and possessed no ability to generate wisdom by themselves. And these implications were inseparably present in their chemical, laboratory and field operations. Wisdom came, if and when it did, from humans.

Patterson, in some unpublished studies carried out with M. Yuasa, R. Aines and G. Kolbasuk during the early 1980's, determined with extreme accuracy and extended meaning the lead contents of trunk wood of various ages in three species of trees. These values were, as usual, orders of magnitude less than concentrations of lead in tree wood reported in hundreds of cases in the scientific literature. So he wrote in reports of studies by his colleagues and himself that their analyzed samples of tree stem wood were from material that constituted a major portion of the earth's biosphere. Since the three sets of values determined in his CIT lab were the only correct ones, the best available estimate for the concentration of lead in the greater part of this gigantic mass of the earth's total biosphere was determined by the mean of only those three sets of numbers.

This instance was one of many such cases generated by Pat and his colleagues analyzing lead in plants, animals, water, and air, all of which ended up proving that literally tens of thousands of concentrations of lead in such materials reported in scientific journals by many other investigators in earlier decades were erroneously high by orders of magnitude. These investigators had attempted to make measurements of microscopic amounts of lead. Such measurements were impossible because of deliberately added as well as inadvertent traces of industrial lead that pervaded, without their understanding nor knowledge, all aspects of their sample collecting, laboratory handling, analytical treatment, and instrument detection procedures. These were lessons hard won by Pat after decades of meticulous work originally developed during his dating of the earth. In Pat's view, so called "researchers" in these cases mistakenly believe that analytical instruments are oracles of wisdom, and that their views and operations typify data-gathering. In his view such synoptic investigation practices have been growing in magnitude in all fields of science, leading to a rapid decline in the integrity of development of scientific knowledge during recent decades.

Pat's rebellious demeanor led him to insult his Caltech colleagues by refusing an appointment to professorship. In his opinion, tenure was an improper and misleading goal for young and future scientists to seek, for it distorted and demeaned the integrity of their thinking as scientists. He would not act, in the views of students, as if he believed otherwise. He was the only person in the history of Caltech who ever refused such an appointment (Figure 4). At the end of his salaried career he apologized and admitted he should have acted more constructively and formally accepted finally an academic appointment to heal old wounds.

However, Pat was always extremely sensitive to the scientific and verbal clamor of the outside world. Thus he took strong measures to keep their noises from intruding upon his thinking. His office is the only one on the Caltech campus that was rebuilt to possess double doors, double walls and ceilings, and double windows. Here he continued up to the end to think and write in complete silence which for four decades instilled in us a continuing uncontaminated heritage for pure knowledge of science and the human world (Figure 5).

Figure 4. Pat Patterson seen unusually attired in cap and gown, as he had shunned any academic appointment for most of his Caltech career. His endearing smile belies a devilish personality and uncompromising scientific spirit.

Figure 5. Pat Patterson receiving the Tyler Prize for Environmental Achievement in 1995, the leading international environment recognition awarded for his exposure of the worldwide insult of lead.

POSTSCRIPT AND ACKNOWLEDGMENTS

Although this manuscript is biographical, it is in fact autobiographical. It was initiated several years ago by the author in a series of interviews, written, and submitted for review by Patterson. In typical style, Pat chose to essentially rewrite it in his own style and language. Thus, except for a few minor editorial changes and a few changes at the time of his death to convert from a present to past biography, it is in fact the story of Pat's life and career in his own words.

EX PLUMBUS UNUM

An Ode to My Mentor: Clair C. Patterson (1922-1995)

When in the course of human events
Comes a life among us so well spent
In matters so heavy, in all ways indeed
The matter was lead in which he did us lead
From the dates of our planets to pollutant curses
The truth was revealed in all ways and courses
From beakers of Teflon boiling late in the night
As the ions of lead in the spectrometer took flight
For the message it told us he tried to explain
Were nanograms of lead coursing through our brains
And so you old plumbus which dated the ages
Is a blight which among us affecting the sages
So adieu, my mentor, in whose steps I did follow
For your shoes are now empty, my feet but to swallow
But he hands me the torch shouting "You did not light its glow!
Here's the light of pure science from ages ago!"

T.M. Church
8 March 1996

CHAPTER 2

SOME REMEMBRANCES FROM ALMOST THREE DECADES OF WORKING WITH PATTERSON

DOROTHY M. SETTLE
Division of Geological and Planetary Sciences
California Institute of Technology
Pasadena, California 91125

This is written on the occasion of the end of an era. Patterson's ultra-clean laboratory era has indeed ended – the old lab is demolished and in its place is a brand new lab, and one can enter that once hallowed ground with street shoes. All that is left now of the clean lab is memories. But what memories they are!

I remember when I first met Pat in 1958. I knocked on his door looking for his technician and was invited to come into his office. Little did I know what awaited me. I was introduced to third degree questioning concerning what I was doing, and was required to explain in detail, complete with theoretical justification, the operation of the Perkin-Elmer flame photometer on which I was doing potassium analyses. I must have passed the test, but that experience was burned indelibly in my mind. Some ten years later, when I was about to begin working in the clean lab, I wondered if I would be grilled every time I entered his office. Those who worked with Pat know the answer was yes.

In 1968 I came to work for Pat, but over the years that changed to working with Pat, and by the time we both retired, a very special relationship had developed in our scientific endeavors. We had reached the point where we were able to finish each other's sentences word for word. To work with Pat was an experience that was challenging, enlightening, and exciting. He used to say that he was a terrible teacher, but when he explained something one-on-one, with a passion derived from his own ideas and knowledge, I don't think anyone could have surpassed him. He taught many things, but above all he taught not to compromise, no matter what. I remember an occasion when he wanted to change a number in an article he was writing for a National Research Council report, and the editor said he was just massaging the data. I had seldom seen him so incensed, and after that I could always get a rise by accusing him of "massaging the data."

How do you put twenty years of working side by side with your best friend on paper? There are so many things I'd like to say that it is almost impossible to choose what to include. We literally traveled from the bottoms of the oceans to the tops of mountains collecting samples which would allow us to measure the extent of world lead pollution. Probably one of my most memorable experiences happened on Enewetak in the Marshall Islands. The SEAREX consortium had built a 20-meter tower there and we worked in shifts around the clock monitoring conditions so that sampling was done only in the proper mode. We went from our lodgings on Enewetak to Sand Island where the tower stood and back again at the end of the shift by small motor boat.

Late one afternoon Pat insisted on staying at the tower beyond the time the day shift ended. He assured everyone that he could handle the boat and that it was absolutely necessary to stay those few extra hours. Naturally I stayed with him and the others crowded into a single boat for the trip back. We finished up whatever it was that had been so important and headed for the beach and the boat. He had some difficulty launching it, and I was more of a hindrance than a help. Voices rose along with tempers until I was on the verge of tears. When the boat was ready to go, I had to call on the walkie-talkie to advise the SEAREX people that we were on the way back. I must have sounded troubled because the person who answered asked me what was wrong and I replied "There is a madman on the beach." Here I must interject that we were on a military base. Women were allowed to come only by special permission because the troops were deprived of female companionship and the commander wanted to avoid any untoward incidents. When asked over the walkie-talkie if I were OK, I replied that I was, and Pat and I headed back to Enewetak. When we arrived at the dock, we were met by the shore patrol, jeeps with blue lights flashing, and just about everyone in the SEAREX contingent. It seems that all of the walkie-talkies were monitored by everyone on Enewetak, and the military thought one of their people had gone berserk on Sand Island. I was mortified, Pat was irate, and I really thought this would be the end of our friendship and my job. He made me go to the Base commander and apologize. For the rest of the time at Enewetak and for some weeks after, little was said between us. Neither of us ever forgot that incident, but we often laughed about it afterwards.

What made Pat so easy to work with, and at the same time so difficult, was that he knew exactly how he wanted everything done, he planned every move with meticulous detail, he did countless dry runs on experiments, and moreover he knew what the answer was going to be before any lab work had been done. For that reason, one of his closest colleagues accused him of having a direct line to God. In the lab he told you how to hold the apparatus, never mind if it were awkward and sometimes painful. You did it that way, and behold, the blanks were the lowest ever! It took us two years to collect, dissect, and analyze one tuna fish from scales to muscle with everything in between, but when we were finished we had something that neither the Bureau of Standards nor the Food and Drug Administration had been able to do. We had a number for tuna muscle that helped delineate the extent of lead pollution in the oceans.

Pat reminded me of a nursery rhyme – there was a little girl who had a curl right in the middle of her forehead, and when she was good she was very very good, but when she was bad she was horrid. I saw both sides, as I'm sure many others also did, but in the final analysis it was more good than horrid. When one remembers Pat, the first thing that comes to mind is his science, which was brilliant, pioneering, and exciting, and I consider myself fortunate to have shared it so closely with him. However, I would like to describe some of the perks I enjoyed while working with him.

When I first came to work with Pat, he made his abhorrence of smoking clear, so rather than argue I stopped smoking, not without great difficulty and much sneaking around to keep him from knowing that I hadn't really stopped. It was not unusual for Pat to go to the lab on weekends, and he often stopped by my home to discuss what he was going to do. One Saturday, as I was in the yard with a hose in one hand and a cigarette in the other, I saw Pat driving down the street. I quickly dropped my cigarette and crushed it out with my foot, forgetting that I was barefoot. It took several days for the blister to heal.

He was a great proponent of exercise and managed to convince me that it was in my best interest to exercise. Three times a week I was at the Caltech track lifting weights and jogging. When I first started I could scarcely walk around the track, but after a few months I could jog two miles, and I was in good enough shape to climb the SEAREX towers and Mt. Etna. I received Mother's Day cards from my cats signed in Pat's handwriting. And when my husband was ill Pat told me that he had moved the phone from Laurie's side of the bed to his so I could call him any time and he would be there. So for me, of equal importance to the science we did together was the deep friendship that developed over 20 years with both Pat and his wife Laurie. They made me feel like a part of the family and I thank them for it. My life was truly enriched by having Pat not only as a colleague but also as a friend.

In 1965 Pat published an article in *Archives of Environmental Health* predicting the degree of lead pollution in people which was received with disbelief and controversy. For the next two decades he worked with his colleagues to prove his theory. When it became clear that humans were indeed contaminated with lead 100 fold above natural levels, he proposed that studies of rats raised in a natural lead-free environment would show how natural biochemical processes operated. This information could then be used to demonstrate harmful effects of lead pollution on people. Musings on why people had poisoned the world with lead brought him to his last interests on the study of human brain functions. We had numerous long discussions concerning differences in brain wiring between scientists and engineers, males and females, orientals and occidentals. Once again, Patterson was on the threshold of pioneering research. He had been pondering this issue for a number of years before he died. I don't understand much of it because it is so complicated that even he couldn't explain it clearly. But I have no doubt that one day in the future it will be shown to be an important area of research.

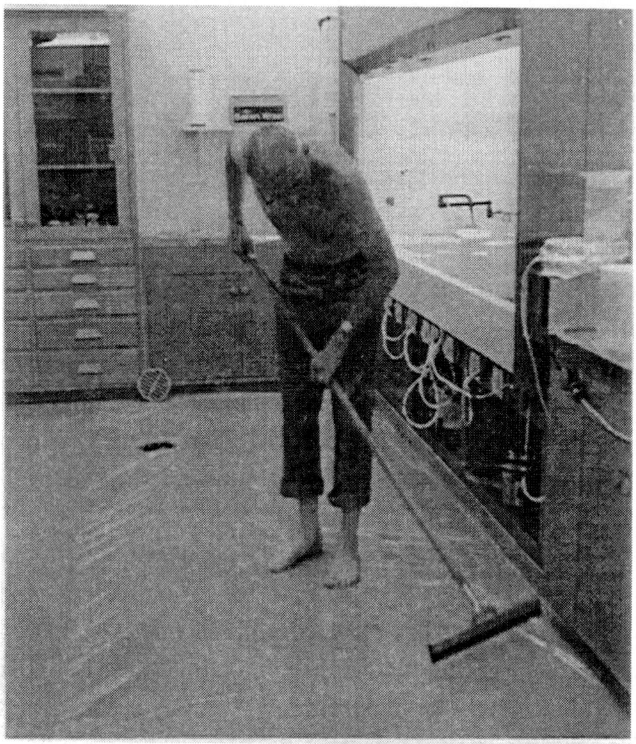

Figure 1. Patterson in the clean laboratory at Caltech.

Figure 2. Patterson and Settle in Hawaii with their air samplers.

Figure 3. Patterson and Settle on the SEAREX Tower in Enewetak, preparing to cover the dry deposition collector and to open the rain collector as a storm approaches.

Figure 4. Patterson in Samoa at the Chief's home in Tula Village.

Figure 5. Patterson at Te Paki Station, North Island, New Zealand.

Figure 6. Patterson on Mt. Etna volcano.

CHAPTER 3

FACTORS INFLUENCING C.C. PATTERSON'S ROLE IN THE EVOLUTION OF THE MULTIDISCIPLINARY FIELD OF BIOGEOCHEMISTRY

A. RUSSELL FLEGAL
Earth Science, WIGS
University of California
Santa Cruz, California 95064

> Some people come to the age of work when there is no work, others when there are wars...Such joining of a man's life with events, large and small, are his unique career... (Hughes, 1971)

It would not be appropriate for me to attempt to chronicle Clair Patterson's evolution as a scientist, because I didn't even meet him until after he had achieved an international renown in the scientific community. However, I have spent two decades trying to determine what factors contributed to that prominence and notoriety; and it seems that some of the same factors contributed to both. These include the truly unique characteristics of Patterson's personality, the exceptional level of support he received from his family and colleagues, and luck. Since others are writing on the first two factors for this book, this essay is limited to a few comments on what I perceive as Patterson's great luck.

Patterson was born to the right parents. They inadvertently gave him good genes for critical thinking, both conceptually and abstractly, that enabled him to develop original scientific concepts rather than pursue the mundane. They also instilled in him an exceptionally strong work ethic, that contributed to the unusual rigor of his studies, and a deep sense of morality, that contributed to the altruistic focus of those studies. Moreover, they provided all of those things without adding the emotional baggage that is so commonly used by individuals in his adopted state of California to rationalize any problems that may occur throughout their lifetime. In any case, Patterson commonly attributed much of his success to the nurturing environment that his parents created.

Patterson was also fortunate in his educational experiences. He grew up in a rural environment in Iowa, where education was emphasized, and then went to a small college, Grinnell, near his home. While many of Grinnell's alumni take pride in it being called the

"Harvard of the Midwest," Patterson took greater pride in being expelled from it. While I never have been able to get a straight answer out of him on the details of that expulsion, his infraction was inconsequential by today's standards. Consequently, the college administrators decided that he deserved a second chance and Grinnell now lists Patterson as one of their most distinguished graduates.

He then began a convoluted scientific career, where the expertise that he developed in one area led to research in another. He first completed an M.S. with George Glockler at the University of Iowa, where he investigated molecular spectra. That expertise qualified Patterson to conduct research on spectrography and mass spectrometry in the Manhattan Project during World War II.

Although World War II and the preceding decade of hard times stunted the development of some of the most gifted individuals in the United States and elsewhere (Elder et al., 1991), it also ushered in a golden age of science. The fields of nuclear chemistry and physics rapidly evolved during that period, largely because their potential applications in warfare were just being recognized. This resulted in programs that brought together some of the most promising young scientists in the United States with some of the greatest chemists and physicists in the world, and provided them with whatever resources were needed in order to produce "the bomb" before the Germans. While it appears that Patterson's involvement in that effort was too late for him to have made any substantive contribution to the development of nuclear weapons, he continued to regret and bemoan his involvement in that program for the last half century of his life.

Patterson's string of luck continued with his subsequent admittance to the doctoral program at the University of Chicago, where much of the seminal research in nuclear chemistry was focused. [Of course, his acceptance at Chicago was a consequence of his exceptional academic record and demonstrated commitment to research.] According to him, there has never been any place like it at that time; and he may be right. Today's textbooks are filled with the discoveries of scientists who were on the faculty at Chicago during that period, and many of the students who came through with Patterson became leaders in their scientific fields.

That exceptional environment was summarized by Patterson (1986) in his brief tribute to Harrison Brown, who was his major professor at the University of Chicago:

"Extraordinary developments that gave rise to much of our present nuclear geochemical knowledge in the earth and planetary sciences were initiated at the University of Chicago after World War II. A young member of the faculty at that time, Harrison Brown, a highly creative scientist fresh from the laboratories of the Manhattan Project, was influential in these developments. World-renowned scientists were among his faculty colleagues in the fields of radiochemistry, inorganic chemistry, organic chemistry, geology, physics, and astrophysics. Those in nuclear geochemistry included Nobel laureate Harold Urey, who with his coworkers was working out the $^{16}O/^{18}O$ method for determining paleotemperatures, stable isotopic fractionations of carbon, and the biochemical origin of life on earth; Nobel laureate Willard Libbey, who with his coworkers was establishing the ^{14}C dating method; Hans Suess, who was working on cosmic abundances of the elements and with Nobel laureate Maria Geppart Mayer was developing models for energy levels in atomic nuclei from these atomic abundances; Gerald Kuiper, who was developing theories for the origin of the solar system; and Mark Inghram, who with coworkers was studying the isotopic and geochronometric occurrences of rare gases in terrestrial and meteoric materials.

"These peers created a challenging environment that enabled Harrison to flourish and realize his full potential as an outstanding scientist who ranks as one of the most stimulating on a world scale. His worthy and exciting ideas and his flair for obtaining positive responses from people influenced both the research aims in a number of fields in nuclear geochemistry, chemical oceanography, and astrophysics and the development of governmental funding for such investigations. He never wanted people to attack solid, reassuring problems infused with an aura of certainty that acceptable solutions would be ground out if the crank were turned. Instead, he enticed people into striving for splendid new views of our world and got them irretrievably committed to such efforts before they began to sense, too late, the costs associated with cantilevering out into the lonely voids of protoknowledge. He automatically dismissed costs as being unworthy of consideration in assessing priorities that might govern directions of research. In this way he served as a distinguished pioneer on the frontier of science, igniting the passions of the unfulfilled, making the unobtainable appear within reach, and thus inaugurating discoveries that, but for him, might still be forgotten."

It is notable that the latter paragraph provides an unusually accurate description of Patterson's role as a mentor. This is somewhat surprising in that Brown's and Patterson's personalities and approaches to science were, according to Patterson, markedly dissimilar. In spite of those reported differences, Patterson emulated his major professor in motivating individuals to exceed their apparent potential. And like Brown, Patterson's success at enticing his colleagues to ignore the considerations of convention, cost, other priorities and time to seek protoknowledge was truly remarkable.

As Patterson described in that tribute, his research with Brown during their years at Chicago and their early years at Caltech focused on the origin of meteorites and their geochronology. The research was based on new techniques in nuclear geochemistry that had been developed as a consequence of the Manhattan Project. Patterson's research has also benefited from collaborations with two more of Brown's exceptional graduate students who were also working on the formation of the solar system, Ed Goldberg and George Tilton. Together they worked on microtechniques to separate small quantities of radioactive minerals developed by Brown's previous associate, Esper Larson, Jr. at Harvard, and they had access to the state-of-the-art mass spectrometer in Mark Inghram's laboratory at the Argonne National Laboratory. With that unique combination of new ideas, new methodologies, and new instrumentation, Patterson and his associates were able to: develop new techniques for trace element analyses in microminerals using isotope dilution mass spectrometry (e.g., Brown and Patterson, 1947); open up the field of uranium-lead dating of common igneous rocks with geochronometric measurements of uranium, thorium, and lead concentrations and isotopic compositions of lead in mineral separates (e.g., Brown et al., 1951); and eventually determine the first accurate ages of meteorites and the earth (e.g., Patterson, 1953; 1956; Patterson et al., 1955).

While providence brought all of those exceptional individuals and incomparable resources together to create an environment where such pioneering studies could occur, they only occurred as a result of the heroic efforts of all of those involved in the research. This is, of course, what ultimately distinguishes exceptional individuals like Patterson from the rest of the field. It was subsequently evidenced to an even greater degree by his, in a sense, singular efforts to quantify the magnitude and sources of lead contamination in the environment. However, Patterson's success in this new area of research was again predicated on a great deal of (bad?) luck.

In that case, Patterson's luck was in discovering that measuring lead concentrations and isotopic compositions in an iron meteorite was not "duck soup" as Harrison Brown had believed it would be (Patterson, 1986). Instead, Patterson found that the problems of lead contamination were enormous, and that essentially all previous reports of lead concentrations in the environment were invalid. Consequently, he had to improve upon the clean-laboratory techniques that George Tilton had developed for measuring milligram quantities of uranium by building an ultra-clean laboratory and developing ultra-clean techniques that reduced the lead blanks in his analyses by orders of magnitude.

Resolving those analytical lead blanks inextricably drew him into studies to quantify and identify the sources of lead contamination in his laboratory. Those efforts inadvertently brought him into studies to quantify and identify the sources of lead contamination in the environment throughout the world. That sequence of events could be considered bad luck by two diverse groups, those who regretted his departure from basic geology and those with vested interests in the lead industry.

Although Patterson's success at dating the Earth is considered one of the principal scientific achievements this century, it is questionable whether he would have enjoyed comparable success in his environmental lead studies if he had not been at Caltech. This is because Caltech had the resources to allow Patterson to conduct his environmental lead studies with minimal distractions. The university provided an unusually large amount of laboratory space for his research (e.g., it is now being divided into two laboratories) and Harrison Brown secured extramural funds to construct the ultra-clean facilities required for that research. The university then created a unique research position for him as "Geochemist." As Caltech's first, last, and only "Geochemist", Patterson had no substantial teaching or administrative obligations that I am aware of, and was allowed to concentrate all of his efforts on his unique research interests.

This relatively unusual arrangement was consistent with Patterson's character. For example, his induction into the National Academy of Sciences was probably delayed for two decades by concerns that he would not accept the invitation. His disregard for honors and titles is illustrated by Patterson's (1986) contribution to the book honoring Harrison Brown (Smith et al., 1986). Several of the other contributors to the book wrote scientific articles, and most of them listed some of their most prestigious titles, honors, and publications in their autobiographies for the book's appendix, About The Contributors. In contrast, Patterson limited his contribution to a few paragraphs about his association with Brown and the following autobiographical statement:

"Clair C. Patterson is a Geochemist in the Division of Geological and Planetary Sciences at the California Institute of Technology, having been a member of the faculty there for thirty-four years. By choice, Professor Patterson has not been active in social affairs in science and has published works in refereed journals."

It is widely rumored that Caltech refused the request of some relatively important donors, who were displeased with the direction of Patterson's research on the sources of environmental lead contamination, to dismiss Patterson from the university. Most other academic institutions might not have been in such a position then, and it is less likely that many institutions would be in such a position now. So, Patterson's luck in timing and location were, once again, impeccable.

After Patterson had determined that essentially all previously published reports on lead concentrations in the environment were erroneously high (e.g., Patterson, 1965), he had the opportunity to rapidly rewrite the literature with a series of publications of accurate lead concentrations in the most accessible environmental matrices. Instead, he elected to focus his efforts on analyses of lead concentrations in the most difficult and inaccessible components of the biosphere. These included ice cores from the polar regions (e.g., Murozumi et al., 1969), sea water collected from the bottom of the ocean (e.g., Schaule and Patterson, 1981), aerosols from the most remote regions of the earth (e.g., Settle and Patterson, 1982), volcanic emissions (Patterson and Settle, 1987), pelagic marine organisms (Settle and Patterson, 1980), blood plasma (Everson and Patterson, 1980), and bones of prehistoric humans (Ericson et al., 1979).

Virtually every one of those analyses and Patterson's discussions of their implications was met with a modest amount of quiet approval from some and an unusual amount of overly loud criticism from many. From my admittedly biased point of view, the approval came from the few scientists who recognized and appreciated the quality and significance of Patterson's research, and the criticisms came from those whose careers and incomes were threatened by the quality and significance of Patterson's research. It is notable that essentially all of Patterson's once controversial analyses have now been substantiated by others who have adopted his ultra-clean techniques. Moreover, Patterson's ultra-clean techniques for environmental analyses have now been promulgated by reports of the National Research Council (National Research Council, 1993) and numerous government agencies.

In addition to substantiating his contrarian nature, Patterson's preliminary analyses of a few selected samples from extreme locations and diverse environmental matrices provided the basis for quantifying the magnitude of lead contamination on a global scale over the past five millennia. Those calculations were based on his belief that it was possible to model anything with a few good numbers and a lot of intuition, as long as the model was based on fundamental scientific principles. In doing so, it appeared to me that the number of his analyses were often inversely related to the length of his discussions of those data. For example, he once rewrote my brief (five paragraphs and one figure) draft of our manuscript on lead in plankton in the northeast Pacific into a fifty page tome with fifteen figures that delineated the global cycle of lead in the marine biosphere throughout the world ocean. Although Patterson's grand extrapolations may have once been questioned, again virtually all of his interpretations of those data have since been corroborated by subsequent analyses.

From my perspective, the field of biogeochemistry was primarily developed by scientists using nuclear techniques, which were developed in the creation of weapons that could destroy the earth, to address evolving concerns about the health of its ecosystem. This seemingly contradictory use of scientific tools was catalyzed by the global radioactive fallout from nuclear explosions. These provided some of the first evidence of the global dispersion of some contaminants and possible fragility of the Earth's ecosystem.

Of course, Patterson was again ideally positioned to apply the skills he had learned from some of the fathers of the bomb to investigate problems associated with the global fallout of industrial lead. This is illustrated by Patterson's concept of normalizing lead concentrations to those of its biogeochemical analogues (calcium and barium) to trace the biogeochemical cycle of lead and its metabolic transport in organisms (Patterson, 1965; Settle and Patterson, 1980). That concept was based on studies of the biogeochemical cycling of ^{90}Sr, which is a product of nuclear explosions and accumulates in bone as a calcium analogue (Elias et al., 1982).

While few are familiar with the normalization of ^{90}Sr, Patterson's "measles figure", which illustrates levels of lead contamination in preindustrial and contemporary humans using lead to calcium ratios, has become an icon. It has been widely reproduced in newspapers and magazine articles, as well as in scientific reports. This includes the recent report by the National Research Council (1993), "Measuring Lead Exposure in Infants, Children, and Other Sensitive Populations," where it is the initial figure. That report is also referred to as "Patterson's Revenge," because it validates everything in Patterson's earlier minority report on lead contamination for the National Research Council.

One area where luck wasn't a factor was Patterson's ability to attract and excite colleagues. He was able to draw a group of dedicated people around him throughout his scientific career. These included the lifelong friends that he acquired during his early research in geochronology and those he acquired during his latter studies of environmental lead contamination. This was evidenced by the diversity of individuals who participated in the symposium, Topics in Geochemistry, honoring Patterson; by the diversity of articles in a special issue of Geochimica et Cosmochimica Acta, which includes some of the presentations from that symposium; and by the diversity of articles in this book. Foremost among all of his friends and colleagues is Dorothy Settle, who has selflessly worked with Patterson throughout most of that time.

The other area where luck doesn't appear to be a factor in Patterson's success is in the support he has received from his wife and family throughout his entire scientific career.

Luck may have brought Pat and Laurie together during their early years in college, but it wasn't what kept them together. Their commitment to each other and their family was rock solid for over half a century. Even with his passing, Patterson's family has continued to support his research and his ideals. In that respect, those of us who are fortunate enough to be part of that extended family are eternally grateful.

This brief essay began with part of a quote from Hughes (1971) that was also used to preface an article by Elder et al. (1991).

"Some people come to the age of work when there is no work, others when there are wars....Such joining of a man's life with events, large and small, are his unique career, and give him many of his personal problems."

Their article, "Talent, History and the Fulfillment of Promise," attributes many of the failures of some of the most gifted (i.e., IQ > 135) individuals in Patterson's generation to realize their potential to the adversities of the Great Depression and World War II. Although this essay may indicate that Patterson's contrasting successes were due to luck, most of it was the fortune he created by thriving on adversity.

ADDENDUM

This essay was written prior to Pat's unanticipated death, with the expectation that he would read it. It was designed to provide another perspective of the factors contributing to Patterson's scientific successes, other than his brilliance, dedication, and drive. It was also designed to emulate the reserved objectivity that Patterson used in his memorial to his mentor, Harrison Brown.

While the editor's request for "anecdotes and interesting stories about knowing and/or working with Dr. Patterson" was left to others, my association with him constitutes the undisputed highlight of my scientific career and one of the highlights of my life. I went to work with him for what was supposed to be a six month postdoc, and remained associated with him for the next two decades. I commonly paralleled this association to a drug addiction, and many of his other colleagues acknowledged that they suffered from the same addiction.

The enthusiasm and commitment to original, rigorous scientific research in his laboratory was impossible for me to escape. As a result, I worked twelve to sixteen hours a day, seven days a week, and wished I had more time to work. I also felt that as hard as I was working, Patterson was working harder, even though Dorothy Settle told me that he would sometimes take naps on his office floor.

Most of that enthusiasm was created by Pat's belief that each sample and each measurement was precious. While I was holding an ice core for Claude Boutron to subsample, Pat pointed to a spot on the core and solemnly pronounced that it was from before the birth of Christ. He then told me that if I dropped the core it would be sacriligious, and that I would be banished from his laboratory for life.

There was also an incredible urgency to get things done. This became immediately apparent to me when I first went to Caltech with the expectation that I would be closely working with Patterson, but found that he remained in his office behind a locked door. After a week, I pounded on his door until he answered and then told him that this was not the association that I expected and that I was leaving. He apologized and explained that he had two types of untreatable cancer and had less than six months to live. Dumbstruck, I anguished all that night about his fate and then asked him early the next day if he had gotten a second opinion. He responded by telling me that he didn't trust doctors and hadn't gotten a first opinion, much less a second opinion, and that he thought that he was going to make it after all. However, he continued to come up with new diagnoses for the next twenty years, and each of them provided a new sense of urgency for whatever project he was working on at the time.

Once Patterson decided that I met his standards, he did everything possible for me. He paid my salary even after I returned to Santa Cruz, so I could be with my family and continue working on projects with him. He then opened his house to our family, because he thought I would be happier and more productive at Caltech if my family was with me. Once, he called my wife in Santa Cruz and told her that Laurie was going to be in Iowa for the summer and that he would move into an apartment there so that my family could live in his house and not have to worry about him getting in the way.

While we didn't take him up on that offer, my family got to know a quite different side of Patterson. When my daughter was two years old, she would proudly state that Uncle Pat had said the Earth was 4.55 billion years old. However, she was more impressed with his ability to pick fleas off his dog and to make ice cream sundaes. Fortunately, or unfortunately, she never saw him square dance. I, in turn, was most impressed with his concerns for both his immediate and his extended families, in spite of his often repeated statements otherwise.

When Dorothy Settle called me before dawn to tell me that Pat had passed away, I told her that I would try to contact his scientific colleagues. I spent three days calling people, and asking them to call people; but I still wasn't able to contact everyone who should have been notified. Everyone that I did contact had personal stories to relate, which were similar to the few I've noted above. The number of people upon whom he had such a profound effect, both scientific and personal, was truly remarkable.

REFERENCES

Brown, H., Inghram, M.C., Esper, Jr., S.L., Patterson, C. and Tilton, G., Isotopic composition of lead and the ages of minerals in a precambrian granite (abstract). *Bulletin of the Geological Society of America*, Vol. 62, p. 1426 (1951).

Brown, H. and Patterson, C., Composition of meteoric matter. I: Composition of the silicate phase of stony meteorites. *Journal of Geology*, Vol. 55, pp. 405-411 (1947).

Elder, G.H., Jr., Pavalko, E.K. and Hastings, T.J., Talent, history, and the fulfillment of promise. *Psychiatry*, Vol. 54, pp. 251-267 (1991).

Elias, R.W., Hirao, Y. and Patterson, C.C., The circumvention of natural biopurification of calcium; along nutrient pathways by atmospheric inputs of industrial lead. *Geochimica et Cosmochimica Acta*, Vol. 46, pp. 2561-2580 (1982).

Ericson, J.E., Shirahata H., and Patterson, C.C., Skeletal concentrations of lead in ancient Peruvians. *The New England Journal of Medicine*, Vol. 300, pp. 946-951 (1979).

Everson, J. and Patterson, C.C., "Ultra-clean" isotope dilution/mass spectrometric analyses for lead in human blood plasma indicate that most reported values are artificially high. *Clinical Chemistry*, Vol. 26, pp. 1603-1607 (1980).

Hughes, E.C., *The Sociological Eye*, Vol. 1, Aldine-Atherton, Chicago, Illinois (1971).

Murozumi, M., Chow, T.J. and Patterson, C., Chemical concentrations of pollutant lead aerosols, terrestrial dusts and sea salts in Greenland and Antarctic snow strata. *Geochimica et Cosmochimica Acta*, Vol. 33, pp. 1247-1294 (1969).

National Research Council, *Measuring Lead Exposure in Infants, Children, and Other Sensitive Populations*. National Academy Press, Washington, D.C., 337 pp. (1993).

Patterson, C.C., *The isotopic composition of meteoritic, basaltic and oceanic leads, and the age of the earth.* Report by the Subcommittee on Nuclear Processes in Geological Settings, National Academy of Sciences/National Research Council, Washington, D.C., pp. 36-40 (1953).

Patterson, C.C., Age of meteorites and the earth. *Geochimica et Cosmochimica Acta*, Vol. 10, pp. 230-237 (1956).

Patterson, C.C., Contaminated and natural lead environments of man. *Archives of Environmental Health*, Vol. 11, pp. 344-360 (1965).

Patterson, C.C., Harrison Brown's influence on twentieth-century developments in geochronology. In: *Earth and the Future: Essays in Honor of Harrison Brown*, K.R. Smith, F. Fesharaki and J.P. Holdren (Eds.), Westview Press, Boulder and London, pp. 10-14 (1986).

Patterson, C.C. and Settle, D.M., Magnitude of lead flux to the atmosphere from volcanoes. *Geochimica et Cosmochimica Acta*, Vol. 51, pp. 675-681 (1987).

Patterson, C.C., Tilton, G. and Inghram, M., Age of the Earth. *Science*, Vol. 121, pp. 69-75 (1955).

Schaule, B.K. and Patterson, C.C., Lead concentrations in the northeast Pacific: evidence for global anthropogenic perturbations. *Earth and Planetary Science Letters*, Vol. 54, pp. 97-116 (1981).

Settle, D.M. and Patterson, C.C., Lead in albacore: guide to lead pollution in Americans. *Science*, Vol. 207, pp. 1167-1176 (1980).

Settle, D.M. and Patterson, C.C., Magnitudes and sources of precipitation and dry deposition fluxes of industrial and natural leads to the North Pacific at Enewetak. *Journal of Geophysical Research*, Vol. 87, pp. 8857-8869 (1982).

Smith, K.R., Fesharaki, F. and Holdren, J.P. (Eds.), *Earth and the Future: Essays in Honor of Harrison Brown*. Westview Press, Boulder and London (1986).

CHAPTER 4

VIGNETTES OF CLAIR PATTERSON

ROBERT P. SHARP
Division of Geological and Planetary Sciences
California Institute of Technology
Pasadena, California 91125

This article consists of brief stories derived from an association with the late Pat Patterson, as friend and colleague, extending over more than 40 years. Some are humorous, some are not, and some have historical content. Most illustrate facets of the complex personality of this most unusual man.

Caltech's Division of Geological Sciences initiated pioneering work in geochemistry in the early 1950s. Our Division Chairman Chester Stock, a noted vertebrate paleontologist, had died of a cerebral stroke, and the staff was considering directions future developments should take. Although southern California is a rich area for mammalian paleontology, Caltech was not the best place for pursuing the subject. Our biologists were devoted to genetics and the emerging field of molecular biology. Most of them didn't know a jack rabbit from a skunk, and their Division Chairman, George Beadle, preferred it that way.

Stock's untimely death stimulated a search for a new endeavor within Geology. Ian Campbell was acting Division Chairman and Al Engel, then a member of our professional staff, had convinced Campbell, without too much trouble I suspect, that we should initiate work in the nascent endeavor of geochemistry. For reasons I don't recall, Engel and I ended up with a commission to explore possibilities. We decided to seek the counsel of Linus Pauling, then chairman of Caltech's Chemistry and Chemical Engineering Division, who had always displayed a warm, supportive interest in the Geology Division. On phoning his office we learned he was home nursing a cold but was accepting phone calls. Not wishing for grass to grow under our feet, we ventured a call to Pauling at home and much to our surprise received an invitation to come up and see him.

As most everyone knows, Linus was interested in crystals and how they were held together. Our division had lots of mineral crystals and this may account in part for his interest in geology. Subsequently, his only daughter, Linda, was to marry a geologist, Barclay Kamb, later to become a Geology Division Chairman, and a good one.

Upon arriving at the Pauling residence, Engel and I found this two-time Nobel laureate, sitting up in bed revising an elementary chemistry text. Most of us would have been conked out, looking at TV, or reading a spicy novel, but not Linus. All this must have been before the virtues of vitamin C had come to his attention. Linus listened to our story about prospects for geochemistry within the Geology Division, thought for a moment or two, and then said, "I think you should look into the possibility of getting a promising young fellow, Harrison Brown, at the University of Chicago to help launch work in the field."

We did some homework on Harrison Brown, he was invited out for a visit, and in due course offered a professorship to come to Caltech and initiate a geochemical program. It was a fortunate choice. Harrison Brown was one of the most innovative, forward looking, entrepreneurial persons I have ever known. He knew where and how to find financial support and had an extremely keen nose for new developments before they blossomed and became recognized by others. When he came west he brought an entourage with him, which included Clair Patterson, Sam Epstein, and Chuck McKinney, among others. This was a rich dowry to bring to any union.

I have long maintained that Harrison's interest in my endeavor had a half-life of about 4-5 years. Thereafter he tended to turn to something new and different. the five years, more or less, that he devoted fully to developing our geochemical program left us a rich heritage, not only in geochemistry, but also in other aspects of our divisional operation. Eventually the focus of his attention shifted to the Foreign Secretaryship of the National Academy of Science. He revolutionized the functions of that office before returning full time to Caltech where he eventually took a position in the Humanities Division and established an operation dealing with world problems of overpopulation, energy, and the like.

It took courage and enthusiasm for exploration on the part of Patterson and Epstein to cast their lot with a bunch of geologists and seismologists who had forgotten most of what they had ever learned of isotopes in elementary chemistry courses. They didn't know us, and we didn't know them, but they were willing to take risks and to venture into the unknown, desirable traits in research scientists. We were further fortunate in that they were interested in isotopic geochemistry, partly along lines originated largely by Harold Urey, then a University of Chicago chemistry professor. Urey had become interested in the moon, which drew him into contact with geologists with whom he seemingly felt comfortable. Remember that Urey was born in Montana and spent some of his early professional years there. Nearly all Montanans are geology friendly because of their surroundings.

Besides requiring mental adjustments on both sides, the ingestion of pure chemists into a geological milieu necessitated significant physical changes. Our two geology buildings of that time, the Mudd and Arms laboratories, were not laboratories in the chemical sense. Major changes involving plumbing, ventilation, and such were required to convert storage rooms, geological laboratories, offices, and other rooms into modern chemical laboratories. Our extensive collection of fossil mammal remains was sold to the

Los Angeles County Museum, and the funds were used to bore holes in the concrete floors of both buildings, driving the occupants wild in the process. Many wished for earplugs, and conversations were restricted to lunch time.

Mass spectrometry was then in its infancy, so we had to build our own instruments and adapt them to our special needs. Fortunately, two talented young jacks of all trades, Victor Nenow and Curtis Bauman, were found, who proved more than equal to the task. To this day the Geology and Planetary Sciences Division has more mass spectrometers than any other operation at Caltech.

Patterson and Epstein were remarkably patient with the lengthy and at times halting process of creating their laboratories. They hung in there and finally could swing into operation. Pat's lab was particularly difficult as he needed an ULTA CLEAN environment in which to work. His PhD thesis at Chicago had dealt with lead-isotope ratios in iron meteorites. That work suffered lead contamination from various extraneous sources. From it Pat learned that he had to have a totally lead-free environment in order to get reliable measurements of the very small quantities of the different lead isotopes in iron meteorites. Very precise rigorous chemistry was required, and no contamination could be tolerated.

In due time we all learned that lead was everywhere: in paint, plumbing facilities, water, metallic instruments, electrical connections, and most of all the smoggy atmosphere of the Los Angeles Basin. "Get the lead out" became almost a Geology Division motto; certainly it was our battle cry. We ultimately claimed with pride that Patterson's lab had the cleanest air in all of southern California.

It was during construction of this laboratory that we all learned what a no-compromise, intense, dedicated, demanding, zealous character we had on our hands in the person of Clair C. Patterson: a satisfactory laboratory would never have been completed without his total dedication to the task.

Fortunately, Harrison Brown didn't need much of a laboratory for himself. His forte was leadership, fund raising, and idea generation. The amount of laboratory work he did was minimal. In early interviews conducted on Brown, one respondent said, "Harrison couldn't pour water from one beaker into another." He didn't have to, he was much more valuable doing other things.

The effort on Patterson's laboratory paid off handsomely. He determined reliable ratios of lead isotopes in earthly rock and most importantly in iron meteorites accurately for the first time and showed that the meteorites originated between 4.5 and 4.6 billion years ago. By extrapolation this became accepted as the time at which the solar planets, including Earth, were also formed. The date has subsequently been confirmed by other radioactive-clock methods. This was a monumental achievement. Had the Crafoord Prize existed at that time, Pat's work would surely have been a strong candidate for recognition.

The experience acquired during laboratory construction and his research on lead isotopes in earth materials led Pat into the field of environmental lead contamination, and the effects of such contamination on humans, especially children. These matters are treated more completely by others writing in this volume, but Pat's work on lead contamination demonstrated compassion for his fellow men, women, and children.

After coming to Caltech, Patterson worked at integrating himself into the strange world of the earth scientists, both socially and intellectually. He soon discovered that field work was an endeavor highly favored by geologists. So he early enlisted in one of our summer field geology camp sessions, then conducted in the Sacramento Mountains of southern New Mexico. This field area lies a little east of the Alamogordo basin and the city of that name, referred to by one of our irreverent graduate students as "Alamogoddamngordo."

This is hot dry country in summer, and the art of transferring the configuration of a contact between two different rock units to a contour map can be trying, besides seeming somewhat esoteric and unnecessary to anyone but a dedicated geologist. The fact that Pat would come out and try his hand at the game, in an effort to be part of his new association, commanded the respect and approval of us all. We learned that Pat was not a half-way guy; he put his heart and soul into every endeavor.

Pat eventually had a field area all his own, one in which he collected specimens to support his study of elemental contamination as passed through a food chain from plants to animals. This did not involve plotting rock contacts on maps, as in New Mexico, but it did involve knowledge and experience in running a field operation.

He had sought help from me in identifying a spot that was remote but not too difficult to access, as free from civilization's contamination as possible. We finally settled on an area in the high Sierra Nevada west of Bridgeport basin, just over the Sierra crest into the eastern, backcountry part of Yosemite Park, accessible only over a foot trail. This was the Kerrick Meadow-Thompson Canyon area, with off-the-beaten-track Thompson Canyon as the point of principal interest. This was no cream-puff operation. A long day's backpack was required just to reach the periphery of the study area, and in winter going was rougher as elevations were around 10,000 feet, and snowshoes were required. This endeavor demonstrated that Pat was indomitable.

Patterson later recognized that some of the best subjects for studying food-chain contamination were fish from the sea. The result was that shortly the third-floor corridor of the Mudd building was lined with refrigerators packed full of frozen fish. Fears that Mudd would soon smell like the Fulton Street fish market proved groundless as Pat handled his fish very carefully out of respect for his colleagues. Those of us with offices off that corridor, however, lived in fear of an extended power failure. Fortunately, it never occurred.

Pat's work on lead contamination brought him into conflict with commercial interests and the scientists employed by certain companies. I can even recall an instance in which one of our Caltech trustees, an executive in a large oil company that added tetraethyl to its gasoline, complained about Patterson's activities to Caltech President Lee Dubridge. DuBridge properly declined to do anything about toning down Patterson's attack on problems of lead pollution. The trustee was one I knew reasonably well. I am confident that he knew better, but was under strong pressures from executives higher up in his company to do something about "that nut at Caltech."

Patterson's conflicts with scientists representing commercial interests motivated him to demonstrate that their measurements of lead contamination in humans were faulty because they did not know what an uncontaminated person was like. Their datum of reference was already contaminated. This started him on a search for possibly uncontaminated humans, during which he managed to get some ancient paleoindian remains from a remote area in South America.

At that time, Patterson and I occupied adjoining offices on the third floor of Mudd Laboratory. We both had a habit of coming to work early, before many people were around. When I was in my office I nearly always left the door open to indicate that I was on board. Early one morning, Pat suddenly appeared in my office door clutching his left hand which was bleeding profusely. I found a towel for him to clasp and rushed him to the emergency room at the nearest hospital. Most of you know that there a patient suffers while registration data such as social security number, health insurance, possibly your great grandmother's maiden name, and other basic information is recorded. In due course, Pat got the attention he needed, but ended up with a scar on his left palm. There are two lessons here. It sometimes pays to come to work early, and Patterson was just as stupid as the rest of us, at times. He was operating on the skull, perhaps the dentition, of his ancient Indian with a dentist's power drill, holding the skull in his hand. The drill slipped and plunged into his hand, lacerating it severely. Luckily, no tendons were cut.

Some years ago Pat's interests in contaminates that passed through the food chain required a trip to South Sea areas. Arrangements for the trip were well advanced when I ran into him one day and said "Welcome to senior citizenship." He immediately fired back, "What da ya mean, senior citizenship?" I said, "Well here you are preparing to go off to the exotic South Seas for several weeks alone with your devoted female laboratory assistant, and no one is upset. Her husband doesn't mind, your wife is unperturbed, and nobody gives a damn. Welcome to senior citizenship." Pat said, "Hm, hm ... (pause) I guess I see what you mean."

Unlike many laboratory scientists, Pat has long believed in maintaining a high level of physical fitness. To that end he once took up distance running, not just jogging but real running. He worked out on the track at La Cañada High school near his home and was soon running, sometimes competitively, with distance runners of their track team. They were surprised that a man of his maturity could not only keep up but give them a run for their money. Pat became so aggressive in this activity that he eventually injured himself by overstraining his legs and had to quit. He succeeded in demonstrating his determination to attain a specific goal, no matter what the cost. Afterwards, when his age began to dictate less demanding forms of exercise, he managed to strain muscles in his back lifting weights. He had a huge supply of determination.

Patterson was not adverse to mixing a little fun along with exercise, and he at one time enjoyed downhill skiing. We made several trips together to ski at Mammoth Lakes in the Sierra Nevada. It was here that, for the only time, I saw Pat at a total loss as to what to do. We had ridden a chairlift well up onto the face of Mammoth Mountain and were preparing to unload. As luck would have it, a girl in the chair ahead of us had stopped smack dab in the narrow path descending from the debarkation point. I managed to slip by on her left side balanced on one ski, but Pat was less fortunate on the right side and managed to trip her up in passing. She fell and put on a performance worthy of a professional Thespian, with much screaming, crying, and commotion. It ultimately developed that she wasn't hurt, but her pride and sensibilities were offended. Pat was devastated and hovered over her apologizing, expressing sorrow, and trying to placate her with endearing terms.

On return from such trips, we usually stopped for dinner in a roadside restaurant. Pat was inclined to talk about his diet on such occasions. He was then, as he was most of his life, slim and trim, and we were inclined to attribute this to burnoff of calories by nervous energy more than by dietary control.

Granted, Pat would eat a meager meal, waving the waitress away when she inquired about dessert. However, when she brought a dessert for others and was about to leave, Pat would slyly ask, "Uh, hum....do you have a piece of that cherry pie out there?" Upon the answer yes, he'd say, "Put a slice in the oven and warm it up." Then as the waitress was departing he would add, "Please put a couple scoops of vanilla ice cream on top." So much for Pat's diet. He was just as human and vulnerable as the rest of us.

Pat complained that noise disrupted his thought processes, so Caltech converted his office into a soundproof room by building a vestibule with two doors separating it from the corridor and installing a lot of acoustical slabs. Consequently, his room was more like a cave than an office, and few of us dared brave the sanctity of that second door, which is probably what Pat desired even more than the lack of noise. Some people generate ideas combing their hair, brushing their teeth, walking to work, shaving, or talking to others, but Pat, like other famous people, preferred to incubate his ideas alone. Some of Pat's ideas were pretty unorthodox, but many were refreshing and right on target. They were numerous and surprisingly varied in scope. He often voiced ideas to me that had nothing to do with science, but concerned such things as the abolition of tenure, a topic on which we happened to see eye to eye. Pat had a wide ranging mind.

Probably few of us would describe Patterson as gregarious, especially in view of his desire for peace and quiet so he could think. Nonetheless, he had an unusual ability to strike up acquaintances with a wide variety of people. His choices were impeccable; for example, Roger Sperry, a neurobiologist and Nobel Laureate, or Saul Bellow, a visiting Fairchild Scholar. Pat was not a recluse. His interests and thoughts spanned a wide spectrum.

Patterson did brilliant research, and I frequently needed to translate it into terms a lay person could understand for the annual report of our division. Pat had a habit, both in speech and in writing, to start in the middle of a subject and work both forwards and backwards simultaneously. Translating him was not an easy task. I would first try to digest his written work, then I would sit and talk with him, seeking answers to things I didn't understand, which were plenty. Eventually, I comprehended and could write a reasonable digest of what he had done or was doing. This brought much credit to our division.

Many of the nice things that happened to me came about because I played on the same team as Clair Patterson. It is a joy to have this opportunity to acknowledge that fact by assembling this account of some shared experiences. Pat was beyond all doubt simply a great guy, whom I regarded with deepest affection.

Check List on some Pattersonian Traits

Trait	Yes	No	Comment
Generous	x		At times to a fault
Predictable		x	Not even with an Ouija board
Free Thinker	x		You bet
Patient		x	Only on occasions
Intelligent	x		Very
Dependable	x	x	Sometimes
Lovable	x		When you knew him well
Forward looking		x	Way out in front
Gentle	x	x	More often than not
Children	x		Loved them
Women	x		Had a way with them
Dedicated	x		Beyond belief
Motivated	x		Strongly
Students	x		All out for a chosen few
Sense of humor	x		Delightful
Wanderlust		x	Except for science
Well dressed		x	Seldom
Kind to animals	x		Consistently
Understood		x	Rarely
Coherent	x	x	Not always
Compassionate	x		Deeply
Intense	x		10^3
Dedicated	x		Totally
Interesting	x		Indeed
Integrity	x		Tops
Courageous	x		Absolutely
Courteous	x		Nearly always
Irascible	x		Occasionally, but not violently

CHAPTER 5

PROFESSOR CLAIR C. PATTERSON, OUR TEACHER, COLLEAGUE, AND FRIEND

KATSUKO SARUHASHI
Executive Director
Geochemistry Research Association
Tokyo, Japan

Clair C. Patterson has had a remarkable impact on the field of geochemistry in Japan. I would like to provide a brief historical statement about how that came to be.

After graduating from the university 50 years ago, I worked under the guidance of Professor Yasuo Miyake. My research involved studying the behavior of CO_2 in the ocean and the distribution profile and behavior of artificial radionuclides between the atmosphere and the ocean.

In 1956, Professor Miyake was invited to Scripps Institute of Oceanography and Caltech. He became acquainted with Professors N.W. Rakestraw, E.D. Goldberg, G Arrhenius, H. Brown, and others including the young scientist C.C. Patterson. Professor Miyake recommended that two of his students, T. Tatsumoto and M. Murozumi, work with Patterson. This resulted in extremely fruitful collaboration: both of these students produced landmark papers on the geochemistry of lead (Tatsumoto and Patterson, 1963; Murozumi et al., 1969).

In 1962, I was studying the distribution of artificial radionuclides in the ocean at Scripps Institute of Oceanography. One day I visited Professor Patterson in his laboratory at Caltech. He kindly welcomed me and escorted me around the campus. He showed me into his laboratory and described his research. Today, 30 years later, I can still remember the excitement I felt listening to him explain the purpose of his work, a discussion which continued as we walked along a tree-lined path when we went out for a coffee break – actually a "tea break" since he didn't like coffee.

In 1966, Professor Murozumi returned to Muroran Institute of Technology from Patterson's laboratory, and established the first ultra clean laboratory in Japan. He also built the first surface ionization mass spectrometer in the country. The cooperation between Patterson and Murozumi continued until Patterson's death in 1995.

Patterson made several visits to our Meteorological Research Institute and Geochemical Laboratory to give lectures and enjoy talks with Miyake and his colleagues.

He highly appreciated our research and once even stated "This is the best geochemical laboratory, No. 1 in the world." His words encouraged our young scientists. On one trip, I escorted him from his hotel to our Institute by the city train, and he asked me "How old is Professor Miyake?" He was surprised by my reply. He suddenly stood up and bowed deeply, right in the middle of the train car, and exclaimed "Oh! Professor Miyake is older than I am by 20 years!" Although I had seen Patterson's severity and concern in the laboratory, I now saw a rather humorous side of this man.

Patterson climbed the Hikada Mountains, Hokkaido, Japan, together with Murozumi and his students. Their purpose was to get samples from virgin areas and determine the impacts of anthropogenic lead on remote ecosystems. They obtained excellent research results which are being used in papers prepared for publication. I should add that Patterson kindly invited several Japanese students to his laboratory to develop their research – S. Nakamura, K. Fujii, K. Igarashi, K. Itoh, and M. Yuasa – who are now all working as associate professors at their respective universities.

Professor Miyake became the President of the Geochemical Society of Japan. He organized the Symposium on Hydrogeochemistry and Biogeochemistry in Tokyo in 1970 which was a great success, and I was fortunate to be able to help with it. Professor Patterson was cordially invited to the symposium as a guest speaker. By that time, of course, he was well-known in Japan for his splendid work on lead in the oceans and in Greenland and Antarctica.

Patterson's lecture at the symposium, entitled "Man's Impacts on the Ocean," was an alarm for severe anthropogenic lead pollution of world-wide environments and ecosystems. After the symposium, other Japanese scientists teamed with Patterson -- for example, excellent work was done by Y. Hirao and H. Shirahata (Hirao et al., 1974; Ericson et al., 1979).

Over the years, Professor Miyake was attacked twice by cancer, sinus cancer over 20 years ago and liver cancer several years later. I remember him in the last days before hospitalization, moving with the aid of a walking stick. During his last ten weeks, spent in the hospital, he prepared several papers on which he included my name as a co-author. He passed away at the age of 82 on October 16, 1990. He was true to his work as a scientist, never wishing to become a high official of the government or university despite a number of offers.

Many years ago, Professor Miyake founded the Goechemistry Research Association to encourage young Japanese scientists to study in this field by presenting the Miyake Prize annually. The 25th Prize was presented in December 1997.

In my heart, the images of both Professor Miyake and Professor Patterson overlap. I hear their questions, such as "What does science mean to you?" and "What do you want to accomplish as a scientist?" In partial answer to these questions, I founded the Association for the Bright Future of Women Scientists when I retired in 1980 as Director of the Geochemical Laboratory, Meteorology Research Institute. To encourage women scientists in their work, and to praise their success in science, I established the Saruhashi Prize.

It is not easy for me to trace the footsteps of these great pioneers. However, I fondly remember my discussions with both Professor Miyake and Professor Patterson.

Professor Patterson indeed had a major impact on geochemistry in Japan. This impact will most likely be felt for many years.

REFERENCES

Ericson, J.E., Shirahata H., and Patterson, C.C., Skeletal concentrations of lead in ancient Peruvians. *The New England Journal of Medicine*, Vol. 300, pp. 946-951 (1979).

Hirao, Y. and Patterson, C.C., Lead aerosol pollution in the High Sierra overrides natural mechanisms which exclude lead from a food chain. *Science*, Vol. 184, pp. 989-992 (1974).

Murozumi, M., Chow, T.J. and Patterson, C., Chemical concentrations of pollutant lead aerosols, terrestrial dusts and sea salts in Greenland and Antarctic snow strata. *Geochimica et Cosmochimica Acta*, Vol. 33, pp. 1247-1294 (1969).

Tatsumoto, M. and Patterson, C.C., Concentrations of common lead in some Atlantic and Mediterranean waters and in snow. *Nature*, Vol. 199, pp. 350-352 (1963).

SECTION 2

SCIENTIFIC CONTRIBUTIONS OF CLAIR PATTERSON AND COLLEAGUES

CHAPTER 6

CLAIR PATTERSON: MORE THAN THIRTY YEARS OF RESEARCH ON ENVIRONMENTAL LEAD

CLIFF I. DAVIDSON
Departments of Civil and Environmental Engineering, and Engineering and Public Policy
Carnegie Mellon University
Pittsburgh, Pennsylvania 15213

INTRODUCTION

Clair Patterson's remarkable research career began with his studies of lead isotopes as a tool in determining the ages of rocks. As described in Chapter 1 by Thomas Church, Patterson succeeded in establishing the age of the oldest rocks on earth, working with other notable geochemists Harrison Brown, George Tilton, and Mark Inghram. This age, approximately 4.5 billion years, has been confirmed by many other investigators.

After this work, Patterson discovered that he could use the methods developed for measuring lead to investigate another important topic: the fate of the vast quantities of lead emitted from anthropogenic activities. It is to this pursuit that he devoted the majority of his career.

In this chapter, we discuss Dr. Patterson's career in three phases. The first phase was the development of his original hypothesis on environmental lead and research on new experimental methods for testing the hypothesis. New methods were necessary because existing techniques were fraught with contamination problems, making accurate lead measurement impossible. The second phase involved using the new techniques to determine concentrations of lead in key environmental samples. Finally, the third phase involved bringing together the new body of data to demonstrate that changes were needed for studying and regulating lead to protect human health. Also included in the third phase were writings on the overall implications of this work for general scientific research.

Although these phases are listed sequentially, they occurred to some extent in parallel. New experimental methods were developed concurrently with the collection and analysis of pristine environmental samples. Similarly, the formulation of new hypotheses and the re-assessment of existing data occurred as new information became available. Each of these three phases will now be described.

PHASE 1: DEVELOPMENT OF THE ORIGINAL HYPOTHESIS AND METHODS TO TEST THE HYPOTHESIS

The statement of the hypothesis and its implications were published in Dr. Patterson's landmark article "Contaminated and Natural Environments of Man" (Patterson, 1965). Table 1, taken from this paper, formed the basis of the original hypothesis.

By considering relative abundances of several trace metals in the earth's crust and in the human body, Patterson hypothesized that natural levels of lead stored in the body of a typical adult would be about 2 mg. He also estimated that adults should ingest about 20 μg of lead per day from food, water, and air in the absence of industrial sources. These values were compared with present-day (1960's) body burdens of about 200 mg of lead, and present-day ingestions averaging about 400 μg of lead per day.

Table 1. Production of some heavy metals in the United States during 1962 (after Patterson, 1965)

Metal	Production (Thousands of Metric Tons)
Lead	1,000
Tin	79
Mercury	2.3
Bismuth	0.9
Germanium	0.02

In the summary of the article, Patterson proposed research to test this hypothesis: it was necessary to determine lead concentrations that had existed in the environment prior to industrial emissions for comparison with present-day concentrations, and to determine the effects of lead at low levels. It was also necessary to evaluate persons responsible for safeguarding public health. The article set the stage for much of the future work in environmental lead measurement as well as health effects research on lead.

The hypothesis that anthropogenic activities were increasing environmental lead concentrations and human exposures by orders of magnitude could not be easily tested. For one thing, it would be necessary to obtain samples that had not been impacted by current lead emissions. Since industrial lead was believed to be reaching virtually everywhere in the civilized world, finding such samples would be difficult indeed. Furthermore, even if such samples could be obtained, it would be a challenge to handle and analyze them without introducing contamination at every step.

Because of the need to minimize contamination, Patterson continued to refine the experimental methods he developed in his early work on lead isotopes. The principle behind these new methods was summarized in a National Bureau of Standards publication by Patterson and Settle (1976): the investigator must know accurately the contribution of lead for each step in the experimental process. These steps include, for example, preparation of the containers, collection of the samples, processing of the samples for analysis, and conducting the analysis. Lead levels in the washwater and in chemical reagents must be known. Table 2, taken from the NBS report, illustrates the extent to which lead contamination must be known.

Table 2. Lead contamination during acid dissolution (after Patterson and Settle, 1976)

		Lead mass (ng)
6 ml HNO_3	0.012 ng Pb x 6 =	0.072
2 ml $HClO_4$	0.018 ng Pb x 2 =	0.036
Dissolution container blank		0.05
Air blank	0.004 ng Pb x 7 =	0.028
Total dissolution blank		0.19

Table 3. Lead concentrations in seawater collected at La Jolla, California in 1972 analyzed by different laboratories, as reported at the 1974 National Science Foundation Workshop (Berand et al., 1976)

Laboratory number	Measurement technique	Measured lead concentration (ng/kg)
1	AS	1300
2	AA	50
3	AA	60
4	AS	600
5	AA	55
6	AS	180
7	AA	120
True Value	ID	14 ± 3

AS = anodic stripping voltammetry
AA = atomic absorption
ID = isotope dilution

Achieving these low levels of lead contamination and developing the ability to measure them requires specialized facilities. The Caltech Geochemistry lab was developed by Patterson for this purpose: it was the first facility of its kind. The lab complex consists of several rooms, each with a distinct function. For example, separate rooms are used for washing sample containers, purifying water and chemical reagents, preparing samples, and analyzing the samples. Water and reagents are purified in specially constructed distillation equipment, and samples are handled with specially designed tools. The entire facility is maintained at positive pressure with purified air. Personnel must enter the lab through a change room, where special clean suits and shoes are provided. Photos of the interior of the lab are shown in Figure 1.

How important are these elaborate procedures? Consider Tables 3 and 4, taken from a 1974 National Science Foundation Workshop (Berand et al., 1976) and from an article by Settle and Patterson (1980), respectively. The "true value" for the seawater samples in Table 3 was determined in the Caltech Geochemistry clean laboratory. Note that all seven of the participating laboratories reported erroneously high readings, illustrating the difficulty of making these measurements without proper procedures.

A similar result is apparent from Table 4: the "other" laboratory has contaminated the sample during analysis by a factor of about 100 over the true concentrations. When the sample preparation as well as analysis are conducted at the other laboratory, the contamination is a factor of 1000 over the true concentrations. The procedures developed by Patterson are clearly necessary to test the original hypothesis.

Table 4. Lead concentrations in tuna (after Settle and Patterson, 1980)

	Measured lead Concentration (ng/g)	
	Analysis at Caltech	Analysis by other lab
Fresh albacore tuna muscle, dissected in Caltech lab	0.3	20
Fresh albacore tuna muscle, dissected in other lab	-	400

Figure 1. Photograph of Clair Patterson in the Geochemistry Clean Laboratory at Caltech.

PHASE 2: DETERMINATION OF LEAD CONCENTRATIONS IN KEY ENVIRONMENTAL SAMPLES

Defining natural lead concentrations required collection of samples removed as far as possible from industrial sources. To this end, Patterson and his colleagues traveled widely, overcoming many obstacles to obtain the samples they needed. Besides personal hardships, each expedition posed new technical challenges during attempts to collect the samples without contaminating them.

We consider here only a small fraction of the many lead measurement projects successfully executed by Patterson. The topics chosen for discussion are lead concentrations in seawater, glacial ice, natural fauna and flora in remote ecosystems, and human bones and teeth.

SEAWATER

Some of the earliest sampling involved collection of ocean water in the early 1960's. Samples were collected from the Pacific Ocean off the southern California coast, the central Atlantic Ocean, the Atlantic Ocean off the coast of Bermuda, and the Mediterranean Sea. Seawater was collected at several depths from the surface down to the sediments in some cases (as deep as 5000 m). Figure 2 shows plots of lead concentration versus depth for three locations (Chow and Patterson, 1966).

The figure shows sharp decreases in concentration with depth below the surface, attributed to deposition of anthropogenic lead. Chow and Patterson used these data to estimate that lead concentrations in deep ocean water at this time were roughly double the expected natural concentrations. If this much lead were indeed being added to the ocean, there should be a great excess of lead in rain and snow even at sites remote from populated areas. To verify this, Tatsumoto and Patterson (1963) measured lead in snow at a remote Northern California site in Lassen Volcano National Park. Results showed that lead in the snow exceeded that expected from soil dust by a factor of 10^4. This excess was attributed to leaded gasoline exhaust.

These early results were confirmed repeatedly in later studies. For example, Schaule and Patterson (1981) and Flegal and Patterson (1983) found substantial gradients of lead in the Central Pacific and Northwest Atlantic Oceans, respectively. In addition, Settle and Patterson (1982) used wet and dry deposition as well as airborne concentration measurements to estimate lead inputs to the North Pacific and North Atlantic Oceans. They found that current inputs are 10-fold greater than those in pre-industrial times. Lead inputs to coastal areas are far more elevated above natural levels than are lead inputs to the open oceans (Patterson et al., 1976). All of these results show that the oceans are contaminated to a significant extent with anthropogenic lead, mostly through atmospheric deposition.

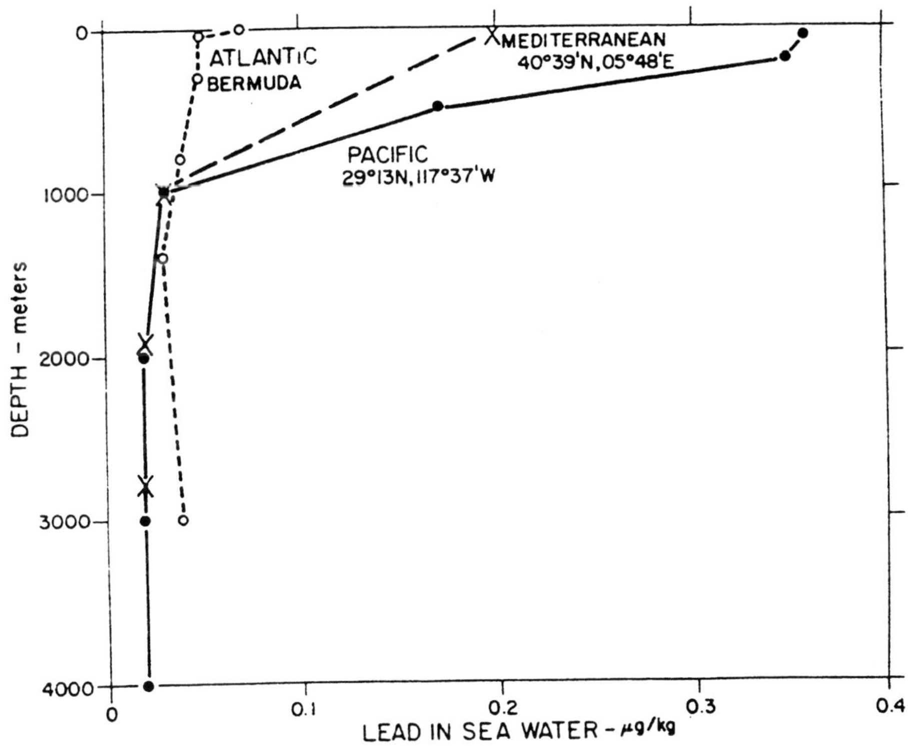

Figure 2. Lead concentration profile in the Atlantic Ocean, Pacific Ocean, and Mediterranean Sea (taken from Chow and Patterson, 1966). (with permission from Elsevier Science)

GLACIAL ICE

Perhaps the best-known of Patterson's studies was his work in Greenland and Antarctica during 1964-67. The polar ice sheets offered what was arguably the best record of changes in environmental lead over time. This is because snow which fell hundreds and even thousands of years ago has remained intact since then, providing a record of past atmospheric conditions.

Accounts of Patterson's expeditions to the Arctic and Antarctic attest to the severe difficulties of this work. Besides having to cope with extreme working conditions, he and his co-workers had to take great precautions to collect uncontaminated samples. The following paragraphs are taken from Patterson and Salvia (1968):

(Referring to work in Greenland:) "We traveled hundreds of kilometers upwind in three fifty-ton caterpillar snow tractors, each pulling a train loaded with supplies, and research and communications equipment. The weather was bad and we traveled through almost continuous storm, checking our course with a sun compass when possible... At a desolate, virgin site, working parties dug a trench fifty feet deep and 300 feet long to collect samples of snowfalls from the last fifteen years.

"In November 1965, I went to Antarctica to conduct similar experiments in the southern hemisphere, accompanied also by three Caltech students and by four students from the University of Canterbury in New Zealand. We flew from New Zealand to McMurdo Sound in Antarctica in Air Force C-130 turbo-prop planes, and from there to the mile-high Byrd Station, halfway between the South Pole and the edge of the Antarctic ice sheet, where even midsummer temperatures fall to ten degrees below zero.

"Here, with electric chain saws and an electric winch, we dug an inclined tunnel more than 300 feet long and 140 feet deep. As in the arctic expedition, all personnel worked ten to twelve hours a day for many weeks, in low temperatures and at high altitudes."

The findings of this work were summarized in the seminal paper "Chemical Concentrations of Pollutant Lead Aerosols, Terrestrial Dusts and Sea Salts in Greenland and Antarctic Snow Strata" (Murozumi et al., 1969). This article had a far-reaching impact. As predicted, the concentration of lead in snow from pre-industrial times was orders of magnitude smaller than current concentrations: < 1 μg/kg in 800 B.C. versus 200 μg/kg in 1965. The value at 800 B.C. represents an upper limit for this sample, based on the extent to which contamination could be minimized. The authors estimate the true concentration to be about 0.4 μg/kg on the basis of terrestrial dust concentrations in the snow. The full record of lead concentrations in Greenland snow is given in Figure 3, one of the most frequently cited graphs in the field of geochemistry.

As explained by Murozumi et al., the graph shows that the lead concentration at the beginning of the European Industrial Revolution (in terms of lead production) was already more than 25 times the natural lead concentration. Steady increases continue up to 1965, with an especially steep rise over the last two decades. Lead in Antarctic snow was considerably smaller than in Greenland, due to the much smaller population of the Southern Hemisphere and the fact that there is relatively little circulation between the hemispheres. It is noteworthy that the general conclusions of Murozumi et al. were confirmed in later work by Ng and Patterson (1981) and Boutron and Patterson (1983, 1987), using greatly improved experimental methods. The results of this later work indicate that Antarctica has the lowest levels of lead contamination on earth: Boutron and Patterson estimate that about 80% of the atmospheric lead over Antarctica is anthropogenic, in contrast to much of the northern hemisphere where this percentage is well over 90%.

Figure 3 is also substantiated by archaeological research conducted by Patterson. He claimed that published values of total lead smelting in past centuries were underestimated because previous investigators had neglected the lead that is smelted during production of silver. When these new amounts of lead are taken into account, total lead emissions correlate well with Figure 3, despite the many assumptions needed to link the two datasets.

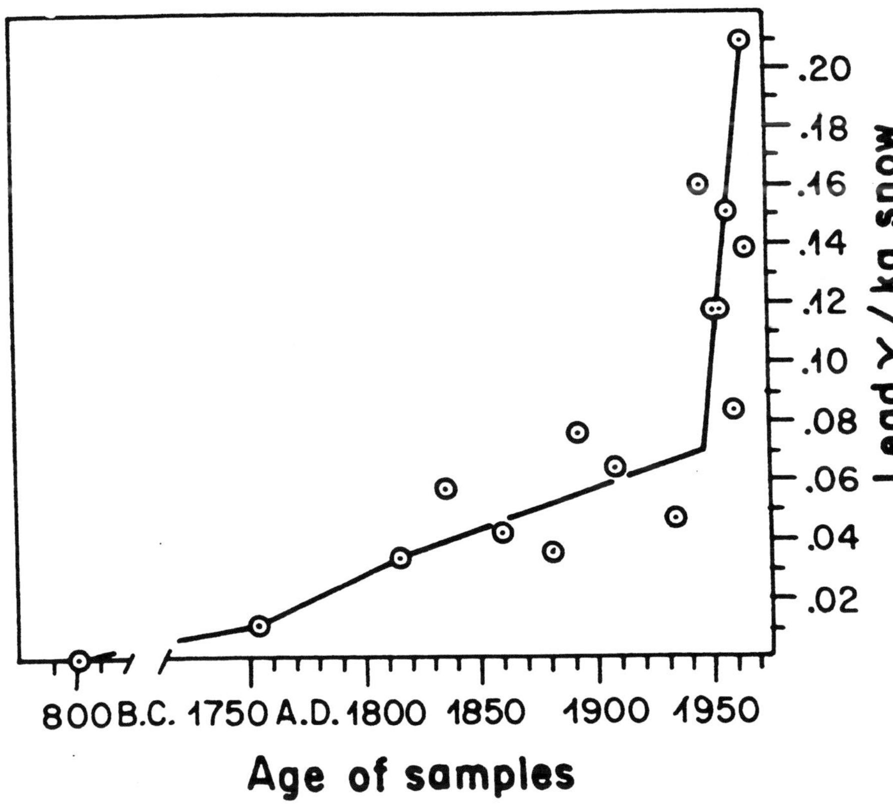

Figure 3. Lead concentration profile in snow strata of northern Greenland (taken from Murozumi et al., 1969).
(with permission from Elsevier Science)

NATURAL FLORA AND FAUNA IN REMOTE ECOSYSTEMS

Patterson's research began with measurements in geochemistry, but studying the impact of world-wide lead emissions also required measurement of lead in plants and animals throughout the food chain. Only through such studies could the distribution and flows of anthropogenic lead in ecosystems be understood.

In the early 1970's, Patterson began work in Thompson Canyon, a remote, high altitude location in Yosemite National Park, California (Hirao and Patterson, 1974). Concentrations of lead and several nutritive metals were measured in soil, soil moisture, vegetation, and several animals (herbivores and carnivores). Dry deposition was measured using surrogate surfaces (nylon), while total input was measured by sampling a snowpack containing eight months accumulation. The flow of lead out of the Canyon was determined by analyzing streamwater and snow runoff.

On the basis of Pb/(K + Ca) ratios, it was estimated that 97% of the aerosol lead entering the Canyon was of anthropogenic origin. Roughly half of this was from stationary sources (mostly smelters), and half was from automobile exhaust. Furthermore, some 98% of the lead entering the Canyon as aerosols remained there (Hirao and Patterson, 1974). Much of this lead continued to be mobile, becoming resuspended into the atmosphere during dry weather and then redepositing onto surfaces in the Canyon (Elias and Davidson, 1980; Davidson and Elias, 1982).

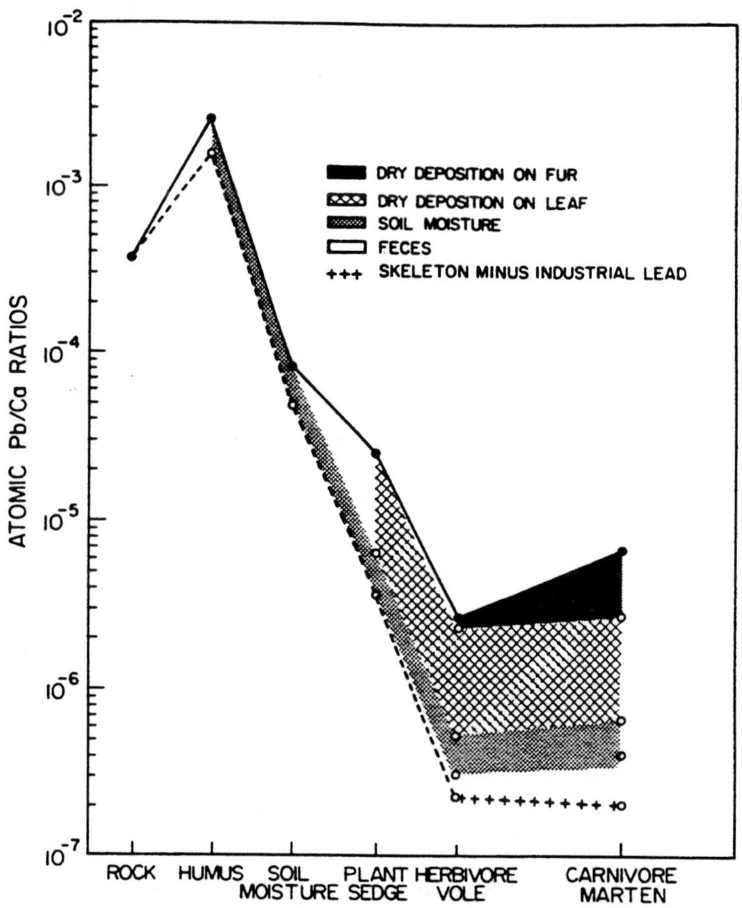

Figure 4. Biopurification of calcium with respect to lead during transfer of calcium in a grazing food chain corrected for industrial lead in soil, plants, and animals (taken from Elias et al., 1982); (with permission from Elsevier Science).

Measurement of lead and nutritive metals in the food chain showed that natural biopurification processes reduce the amount of natural lead transferred to successive trophic levels. For example, values of the atomic Pb/Ca ratio in vegetation are less than those in soil moisture and humus; similarly, values of this ratio in herbivores (voles, a species of meadow mouse) are smaller than those in vegetation. The input of industrial lead can circumvent this biopurification, however. Anthropogenic aerosol lead deposits on the leaves of vegetation, overriding the biopurification that occurs when soil moisture is taken up through the plant roots. Similarly, aerosol lead deposits on the fur of animals and is ingested when the animals groom themselves. Ratios of Pb/Ca in environmental samples representing the various trophic levels are shown in Figure 4 (Elias et al., 1982).

The lower curve in the figure illustrates that natural values of Pb/Ca decrease as these elements progress through the food chain. This decrease is attributed to processes that inhibit the transfer of toxic lead while allowing nutritive calcium to move on to higher trophic levels. The shaded regions indicate lead that has been added to the ecosystem by atmospheric input. Finally, the upper curve shows overall current levels of Pb/Ca that have been altered by this input: the impact of anthropogenic lead is greatest at the highest trophic levels.

These samples were also analyzed for barium, a toxic metal which behaves similarly to lead in terms of uptake by organisms in an ecosystem. Ratios of Ba/Ca decrease as expected in successive trophic levels due to natural biopurification processes. The data for barium, which has very few anthropogenic emissions, thus serve as a control: the Pb/Ca ratios in Thompson Canyon must be due to anthropogenic input. Overall, these results demonstrate that lead pollution has infiltrated throughout the food chain and has altered the flow of lead in this remote terrestrial ecosystem.

Concurrently with this work, Patterson was also conducting measurements in remote marine ecosystems. For example, the work of Burnett and Patterson (1980) showed that the natural biopurification processes in certain marine plants and animals (e.g. kelp and invertebrate herbivores) are greater than those in the terrestrial ecosystem studied.

Of particular interest was measurement of lead in tuna fish. Table 4 above showed the difficulty of obtaining accurate data for lead at low concentrations, using tuna as an example (Settle and Patterson, 1980). This same study also showed that large errors existed in published lead concentrations in food due to contamination in the laboratories of other investigators. Through accurate concentration measurements in processed and unprocessed tuna, Settle and Patterson showed that food processing and the use of lead-soldered cans can elevate lead concentrations by a factor of 10^4 over the amount occurring naturally in the tuna.

Human Bones and Teeth

Following the ecosystem studies, Patterson hypothesized that natural lead levels in humans could be calculated from the biopurification factors estimated earlier. The data from Thompson Canyon showed that Pb/Ca is reduced by 5×10^{-4} between the first trophic level (crustal rock) and the last (carnivores). Using the average value of 6.4×10^{-5} for Pb/Ca in North American rocks, he estimated that the human body should have a ratio of $\sim 2 \times 10^{-8}$ in natural conditions.

Patterson set out to investigate this issue in the late 1970's (Ericson et al., 1979; Patterson et al., 1987). He and his co-workers obtained pre-industrial human bones and teeth dating back as much as 4500 years, which they analyzed for lead, barium, and calcium concentrations. Data for similar samples from present-day humans were taken from the literature. As in the ecosystem studies, barium was used as a control since natural levels of barium would be expected in both pre-industrial and modern samples. Results are shown in Table 5, taken from Ericson et al. (1979).

The table shows that values of Pb/Ca in the pre-industrial samples are much lower than those in modern samples, while values for Ba/Ca are comparable in both datasets. The authors state that several of these pre-industrial samples have been contaminated by soil moisture lead as evidenced by slightly elevated Ba/Ca ratios, and therefore should not be considered reliable. The value of Pb/Ca in the Egyptian sample, 70×10^{-8}, is believed to reflect widespread lead smelting in the Mediterranean region at that time. When the unreliable data are discarded and the remaining data are corrected for age, the best value of Pb/Ca in pre-industrial Peruvian bones is about 6×10^{-8}, roughly 1/500 of the value observed in modern humans.

Table 5. Ratios of Pb/Ca and Ba/Ca in teeth and bones of pre-industrial and present-day humans. Data for present-day samples are taken from the literature (see Ericson et al., 1979 for citations).

Individual	Atomic Age, years before present	Atomic Sample type	Pb/Ca (Multiply values by 10^{-8})	Ba/Ca (Multiply values by 10^{-6})
Peruvian A	3000-4500	Tooth	5.7	1.5
		Mandible	57.	15.
B	2500-3000	Tooth	16.	3.0
		Femur	77.	9.
C	1400-2000	Tooth	2.3	1.3
		Mandible	12.	2.2
D		Mandible	9.	2.8
E		Mandible	200.	20.
F		Tooth	3.7	2.9
Egyptian	2200	Vertebra	70.	0.8
American	Present	Tooth	2000.	2.9
American	Present	Skeleton	3500.	2.9
Briton	Present	Skeleton	2100.	7.0

Patterson also considered lead concentrations in human blood. Using extreme precautions for minimizing contamination, he and his co-workers measured lead in plasma and whole blood of two individuals (Everson and Patterson, 1980). One of these had excessive occupational exposure to lead, while the other had a more typical exposure. Results showed that the plasma lead concentration in the former individual was much greater than that in the latter, as expected. However, both concentrations were considerably smaller than plasma lead levels reported by other investigators for subjects with a wide range of exposures. This discrepancy was attributed to lack of sufficient contamination control by these other investigators. Lead concentrations in whole blood measured by Everson and Patterson were comparable to those reported in other studies. The authors conclude that strict contamination control is needed for reliable measurement of lead in plasma due to the low concentrations expected.

PHASE 3. IMPLICATIONS AND SIGNIFICANCE OF THE WORK

Patterson summarized the significance of much of this work in a chapter of the National Academy of Sciences Report on Lead in the Human Environment (Patterson, 1980). The findings show consistently that anthropogenic activities have redistributed lead throughout the environment. Concentrations of lead in urban areas are several orders of magnitude greater than they were in natural conditions, and even the most remote areas of the world contain lead from industrial activities. This has resulted in exposures that are far greater than those at the time when living organisms evolved. Such discoveries have provided a basis for new research on human exposure and health effects of lead at low levels (e.g., Needleman, 1980; Rutter and Jones, 1983; Rosen, 1985). The results also demonstrate the need for fundamental research on biochemical changes in living cells in the presence and absence of lead.

The work of Clair Patterson goes beyond these issues, however. In his acceptance speech for the V.M. Goldschmidt Medal of the Geochemical Society, Patterson discussed two opposing philosophies of scientific research (Patterson, 1981). Utilitarian research is based on the need of a species to modify the environment in order to survive and flourish. The goal of the research is to gain the knowledge needed to carry out these modifications. Classical scientific research, on the other hand, is motivated by the intrinsic worth of knowledge. The classical scientists have much in common with artists and other humanists, in that their work reflects foremost a high regard for human life and dignity. Having this perspective, classical scientists use their creative talents to generate knowledge about themselves and the world around them.

In the words of Patterson, "Environmental lead studies in a classical scientific mode were uniquely situated to give a true measure of this new utilitarian kind of science." In short, utilitarian research provided the technology that resulted in a lead pollution problem of global proportions. Furthermore, attempts to understand the magnitude of the problem were thwarted by additional utilitarian research that produced erroneous data and incorrect conclusions. Classical scientific research has removed the uncertainties associated with the extent of lead pollution, and has served as a basis for planning further research on the effects of lead. The work has also provided a basis for new regulations that can better protect public health.

In light of the above arguments, perhaps the most important implication of Patterson's work is a new understanding of the dangers of misguided science. There are often surprises when new technologies are introduced; these surprises can become catastrophes if the technologies are applied on a large scale without appropriate precautions.

Dr. Patterson has provided scientists everywhere with a clear message: the complexity of the world and our lack of understanding about it demands that we foster a great respect for natural living systems. We must keep this respect first and foremost as we select questions for research and attempt to answer them. Only through such efforts can we avoid repeating the mistakes of the past.

SUMMARY

The career of Dr. Clair Patterson has been discussed in terms of three phases of work. These are described sequentially, although the work in the three phases took place to some extent in parallel.

First, the original hypothesis was formulated and methods to test it were developed. The essence of this hypothesis was that anthropogenic activities had resulted in redistributing lead throughout the world, producing environmental concentrations and human exposures far in excess of those occurring naturally. Testing this hypothesis required estimates of lead concentrations and exposures both currently and in natural, pre-industrial times; the latter could be estimated from measurements in remote areas, in glacial ice thousands of years old, and in pre-industrial human remains. The measurements required specialized techniques to minimize contamination caused by anthropogenic lead that had become ubiquitous. For this purpose, Patterson designed the Caltech Geochemistry clean laboratory, complete with custom-made tools.

The second phase involved making the measurements. Four types of samples are discussed here: seawater, glacial ice, flora and fauna in remote ecosystems, and human bones and teeth. In each case, concentrations of lead were consistent with the original hypothesis.

Seawater samples obtained at various depths from the Atlantic and Pacific Oceans showed lead concentrations that were highest at the surface, decreasing with depth. Samples of glacial ice showed that lead levels in modern Greenland snow were much greater than snow that fell hundreds of years ago, with even greater differences when compared with ice over 2000 years old. Plants and animals in both terrestrial and marine ecosystems were found to contain anthropogenic lead; it was discovered that this lead could override natural processes that help to reduce the amounts of lead moving through the food chain. Finally, human bones and teeth over 1000 years old were found to contain far less lead than those from present-day humans.

The third phase of work involved exploring the significance of the results. Certainly, the findings were and still are important in guiding future research on the effects of lead, especially human health effects. The findings also served as a basis for new regulations limiting the use and emissions of lead. However, the work also pointed out a serious flaw in the way new technologies are introduced: the damage was considerable before we recognized it and began to reverse the trends.

Dr. Patterson's work stands out as one of the great scientific achievements of the present day. Through unfaltering determination, he devoted his career to establishing the true extent of a serious environmental and public health problem. He gave us a legacy that will guide our work and those of future scientists for years to come.

REFERENCES

Berand, P., Bewers, J.M., Duce, R., Fitzgerald, W.F., Hartling, S., Hunt, C. D., Patterson, C., Settle, D., Smith, R., Walsh, P., Windom, H. and Wong, C.S., Comparison determinations of lead by investigators in both their home laboratory and in an isotope dilution laboratory. *Marine Chemistry*, Vol. 4, pp. 389-392 (1976).

Boutron, C.F. and Patterson, C.C., The occurrence of lead in Antarctic recent snow, firn deposited over the last two centuries and prehistoric ice. *Geochimica et Cosmochimica Acta,* Vol. 47, pages 1355-1368 (1983).

Boutron, C.F. and Patterson, C.C., Relative levels of natural and anthropogenic lead in recent Antarctic snow. *Journal of Geophysical Research*, Vol. 92, pp. 8454-8464 (1987).

Burnett, M.W. and Patterson, C.C., Perturbations of natural lead transport in nutrient calcium pathways of marine ecosystems by industrial lead. *Isotope Marine Chemistry*, E.D. Goldberg, Y. Horibe and K. Saruhashi, editors, Uchida Rokakuho Publishing Co. Ltd., pp. 413-438 (1980).

Chow, T.J. and Patterson, C.C., Concentration profiles of barium and lead in Atlantic waters off Bermuda. *Earth and Planetary Science Letters*, Vol. 1, pp. 397-400 (1966).

Davidson, C.I. and Elias, R.W., Dry deposition and resuspension of trace elements in the remote High Sierra. *Geophysical Research Letters*, Vol. 9, pp. 91-93 (1982).

Elias, R.W. and Davidson, C.I., Mechanisms of trace element deposition from the free atmosphere to surfaces in a remote High Sierra canyon, *Atmospheric Environment*, Vol. 14, pp. 1427-1432 (1980).

Elias, R.W., Hirao, Y. and Patterson, C.C., The circumvention of natural biopurification of calcium along nutrient pathways by atmospheric inputs of industrial lead. *Geochimica et Cosmochimica Acta*, Vol. 46, pp. 2561-2580 (1982).

Ericson, J.E., Shirahata H., and Patterson, C.C., Skeletal concentrations of lead in ancient Peruvians. *The New England Journal of Medicine*, Vol. 300, pp. 946-951 (1979).

Everson, J. and Patterson, C.C., "Ultra-clean" isotope dilution/mass spectrometric analyses for lead in human blood plasma indicate that most reported values are artificially high. *Clinical Chemistry*, Vol. 26, pp. 1603-1607 (1980).

Flegal, A.R. and Patterson, C.C., Vertical concentration profiles of lead in the Central Pacific at 15°N and 20°S. *Earth and Planetary Sciences Letters*, Vol. 64, pp. 19-31 (1983).

Hirao, Y. and Patterson, C.C., Lead aerosol pollution in the High Sierra overrides natural mechanisms which exclude lead from a food chain. *Science*, Vol. 184, pp. 989-992 (1974).

Murozumi, M., Chow, T.J. and Patterson, C., Chemical concentrations of pollutant lead aerosols, terrestrial dusts and sea salts in Greenland and Antarctic snow strata. *Geochimica et Cosmochimica Acta*, Vol. 33, pp. 1247-1294 (1969).

Needleman, H.L. (Ed.), *Low Level Exposure*, Raven, New York, (1980).

Ng, A., and Patterson, C.C., Natural concentrations of lead in ancient Arctic and Antarctic ice. *Geochimica et Cosmochimica Acta*, Vol. 45, pp. 2109-2121, (1981).

Patterson, C.C., Contaminated and natural lead environments of man. *Archives of Environmental Health*, Vol. 11, pp. 344-360 (1965).

Patterson, C.C., An alternative perspective -- Lead pollution in the human environment: origin, extent and significance. *Lead in the Human Environment*, National Academy of Sciences/National Research Council, National Academy Press, Washington, D.C., pp. 265-349 (1980).

Patterson, C.C., Acceptance speech for the V.M. Goldschmidt Medal. *Geochimica et Cosmochimica Acta*, Vol. 45, pp. 1383-1388 (1981).

Patterson, C.C. and Salvia, J.D., Lead in the modern environment. How much is natural? *Scientist and Citizen*, Vol. 10, pp. 66-79 (1968).

Patterson, C.C. and Settle, D.M., The reduction of order of magnitude errors in lead analyses of biological materials and natural waters by evaluating and controlling the extent and sources of industrial lead contamination introduced during sample collecting, handling, and analysis. Accuracy in *Trace Analysis: Sampling, Sample Handling, and Analysis*, P. LaFleur (Ed.), National Bureau of Standards Special Publication 422, pp. 321-351 (1976).

Patterson, C., Settle, D. and Glover, B., Analysis of lead in polluted coastal seawater. *Marine Chemistry*, Vol. 4, pp. 305-319 (1976).

Patterson, C.C., Shirahata, H. and Ericson, J.E., Lead in ancient human bones and its relevance to historical developments of social problems with lead. *Science of the Total Environment*, Vol. 61, pp. 167-200 (1987).

Rosen, J., Metabolic and cellular effects of lead: a guide to low lead toxicity in children. In: *Dietary and Environmental Exposure to Lead*, K. R. Mahaffey (Ed.), Elsevier/North-Holland Biomedical Press, Amsterdam, pp. 157-185 (1985).

Rutter, M. and Jones, R.R. (Eds.), *Lead versus Health*, John Wiley and Sons, Ltd., Chichester (1983).

Schaule, B.K. and Patterson, C.C., Lead concentrations in the northeast Pacific: evidence for global anthropogenic perturbations. *Earth and Planetary Science Letters*, Vol. 54, pp. 97-116 (1981).

Settle, D.M. and Patterson, C.C., Lead in albacore: guide to lead pollution in Americans. *Science*, Vol. 207, pp. 1167-1176 (1980).

Settle, D.M. and Patterson, C.C., Magnitudes and sources of precipitation and dry deposition fluxes of industrial and natural leads to the North Pacific at Enewetak. *Journal of Geophysical Research*, Vol. 87, pp. 8857-8869 (1982).

Tatsumoto, M. and Patterson, C.C., Concentrations of common lead in some Atlantic and Mediterranean waters and in snow. *Nature*, Vol. 199, pp. 350-352 (1963).

CHAPTER 7

AUTOMOTIVE LEAD POLLUTION: CLAIR PATTERSON'S ROLE IN STOPPING IT

JEROME O. NRIAGU
Department of Environmental & Industrial Health
School of Public Health
The University of Michigan
Ann Arbor, Michigan 48109

INTRODUCTION

The discovery of the antiknock properties of tetraethyl lead (TEL) in the early 1920's created a storm of controversy over the safety of using a known toxin as a gasoline additive. The high incidence of industrial lead poisoning, including many deaths, in the production and distribution facilities served to heighten the concern about the potential public health effects of the lead in automotive exhaust. With an effective industrial lobby and a mindset of industrial progress, such concerns were stifled and leaded gasoline went on the market without any assessment of the public health risks. To forestall costly regulatory controls and any damaging negative reports, it became expedient to establish a sophisticated strategy and detailed apparatus which effectively secured a hegemonic control on scientific information on environmental lead hazards. As the industrial perspective on public health risks remained unquestioned, leaded gasoline enjoyed an unfettered growth for nearly five decades. This report describes Clair Patterson's leadership role in challenging the unique and unparalleled hegemony on scientific knowledge on the dangers of environmental lead poisoning. The lifting of the veil of self-serving interests encouraged poignant clinical and environmental studies which left no doubt that tetraethyl lead additive was, in fact, a curse of the gods rather than a "gift from God".

THE DEBUT OF LEADED GASOLINE

Charles F. Kettering, a true giant of industrial development in the United States, had every reason to detest the engine knock. He was accused of "inventing" and perpetrating the engine knock, in reference to the misbehavior of a 1912 model Cadillac which was the first high compression car to be equipped with battery ignition (self-starter), a Kettering innovation (Robert, 1983). His hatred for the knock was intensified when his Delco-Light engine, a small 32-volt farm lighting system, emitted such an ear-piercing knock that Kettering was forced by fire underwriters to reduce the compression and hence the power of the engine. With a zeal and an indefatigable spirit characteristic of a true American pioneer, he set out to overcome his nemesis, the traumatic knock.

In 1916, Kettering assembled a team in his Dayton Research Laboratory and charged it with (a) discovering the real cause of the knock, (b) finding a satisfactory antiknock additive, and (c) discovering a feasible way to manufacture the antiknock agent. The antiknock properties of tetraethyl lead were discovered on December 9, 1922. The Kraus-Callis process for manufacturing tetraethyl lead was patented in the same year, and Ethyl Corporation was created in August 1924 to make and sell leaded gasoline. Details of these events have been well chronicled (Nickerson, 1954; Robert, 1983; Nriagu, 1990). It should be said that Ethyl Corporation, with its monopoly on TEL production, was a monument to Kettering's pioneering spirit -- he conceived of the need for such an enterprise, fashioned its destiny, and probably wrote the first advertisement on its product (Figure 1).

The first gallon of leaded gasoline was sold to a venturesome motorist by the Refiners Oil Company station at Sixth and Main streets, Dayton, Ohio on February 1, 1923. Less than four months later, the new fuel received worldwide publicity when the first three finishers in the Indy 500 motor race were powered by leaded gasoline. The number of ethylizers (the hand pumps for dispensing the leaded gasoline) increased from about 30 in August 1923 to 1200 in May 1924 and exceeded 17000 by October of the same year (Nriagu, 1990). By May 1925, well over 300 million gallons of leaded gasoline had been sold (USPHS, 1925). The rosy picture quickly became clouded, however.

Even before any widespread commercialization of leaded gasoline, many scientists, public health experts, labor and corporate leaders, and government officials were aware of the dangers posed by lead in automotive exhaust (Rosner and Markowitz, 1985; Graebner, 1988). Thomas Midgely Jr., credited with discovering the antiknock properties of TEL, had to go to Florida in February 1923 for rest and recuperation from the effects of too much exposure during the several months when he experimented with the compound. Public opinion became galvanized when large numbers of workers in the experimental laboratory as well as the TEL manufacturing and handling facilities were either killed or came down with severe palsies, tremors, hallucinations and other serious neurological symptoms of organic lead poisoning. The plants were promptly dubbed the "house of butterflies" (because many workers hallucinated about insects during their bouts of lead poisoning) and the leaded gasoline was often referred to as "loony gas" (Rosner and Markowitz, 1985). The following excerpt from a report by Alice Hamilton captures the essence of the early concerns about automotive lead poisoning:

"A new problem of public health and of industrial hygiene has suddenly arisen in this country as a result of the addition of tetra-ethyl lead to motor car gasoline. This procedure was adopted about two years ago, but it did not come to the knowledge of the

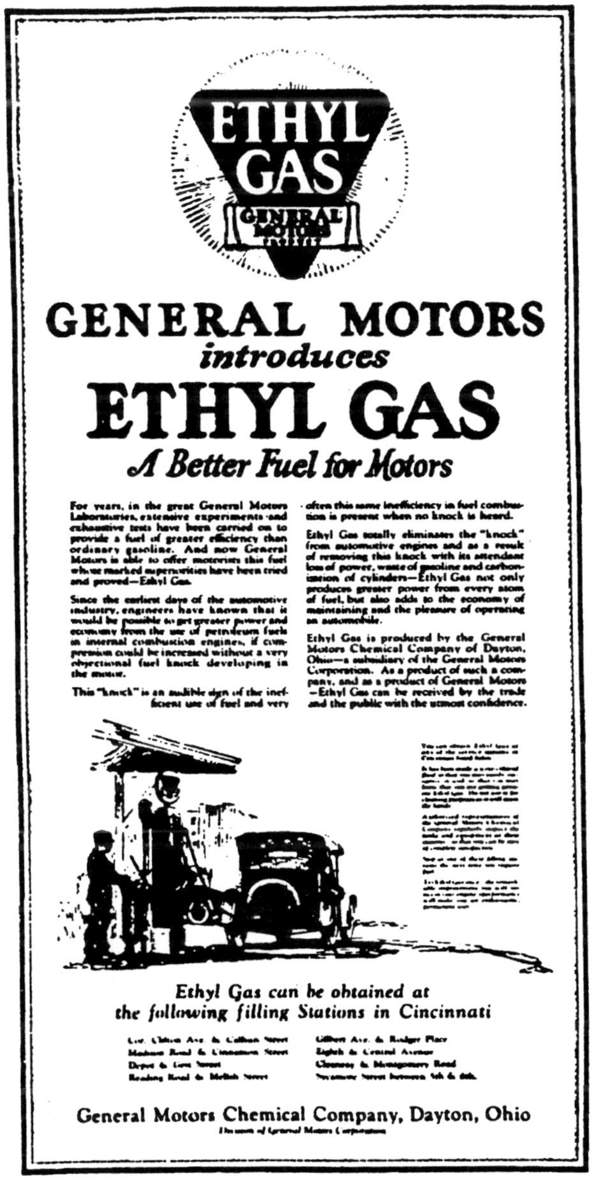

Figure 1. First advertisement of leaded gasoline published in the Cincinnati newspapers during the summer of 1923. The name, "*Ethyl Gasoline*" hides the fact that it contains lead (from Robert, 1983).

general public until last fall when a number of men were poisoned in a New Jersey plant and five died. The distressing details of these cases of insanity and death and the unknown nature of the new poison aroused much excitement and alarm, and induced the public health officials of New York City and the states of New York and New Jersey to forbid the use of "ethyl gasoline" on the ground of danger to the public from poison contained in exhaust gases of motor engines burning such gasoline. This prohibition has recently been rescinded in the two states but is still in force in New York City" (Hamilton et al., 1925, p. 1481).

In an unprecedented move which turned out to be a major public relations coup d'etat, Ethyl Corporation voluntarily withdrew leaded gasoline from the marketplace and persuaded the Surgeon General of the United States Public Health Service to convene a conference of eminent persons in an attempt to resolve the controversy. The agenda and format for the conference were determined by companies with vested interest in the continuing production of TEL. Frank Howard, a spokesman for the Ethyl Gasoline Corporation succinctly summarized the industry's position at the meeting:

"We are engaged in the General Motors Corporation in the manufacture of automobiles, and in the Standard Oil Company in the manufacture and refining of oil. On these things our present industrial civilization is supposed to depend. Now as a result of 10 years' research on the part of the General Motors Corporation and 5 years' research by the Standard Oil Co. we have this apparent gift of God — of 3 cubic centimeters of tetraethyl lead — which can be produced at a low figure, an inconsequential figure ultimately, and which will permit that gallon of gasoline ..... to go perhaps 50 percent further, and if our optimistic engineering friends are correct will make it go 100 percent further in the long run. We are presented with that gift of God which enables us to do that" (USPHS, 1925, p. 106).

Robert Kehoe (more about him later) noted that "when a material is found to be of this importance for the conservation of fuel and for increasing the efficiency of the automobile, it is not a thing to be thrown into the discard on the basis of opinion. It is a thing which should be treated solely on the basis of fact" (USPHS, 1925, p. 70). The demand by the industries for data and scientific proof for harm to the public at large was compelling. At the time, there was little information about the levels or possible long-term effects of automotive lead in the environment. Under the circumstances, the concerns about widespread use of the much heralded powerful antiknock compound were dismissed with the tacit understanding that great innovations involve some risk. The long-term studies recommended by the Blue Ribbon Panel set up by the Surgeon General (USPHS, 1925) were never conducted. The resolution of the controversy surrounding the debut of leaded gasoline was a clear victory for an occupational as opposed to environmental perspective on lead pollution. The risks to the general public raised by Alice Hamilton, Yandell Henderson and others during the 1920's (USPHS, 1925; Rosner and Markowitz, 1985) were ignored for nearly 40 years when they were taken up again with vigor and unparalleled force of conviction by Clair C. Patterson.

HEGEMONY ON SCIENTIFIC KNOWLEDGE ABOUT LEAD POISONING

> "The bulk of the investigative work which bears on lead in the general environment of man has been carried out, under financial sponsorship of the manufacturers and distributors of lead containing antiknock compounds in the United States" (Robert Kehoe, Committee on Public Works, 1966, p. 207).

In view of the storm generated by the "house of butterflies", Ethyl Corporation, General Motors, the Kettering Foundation, and the lead industry associations sought to forestall any negative publicity and any costly legislation pertaining to the production, distribution and use of leaded gasoline. They banded together to create a body of scientific opinion that was ideally suited to the manufacturers of TEL and automobiles by doing or underwriting most of the significant lead-related scientific research. A key architect in this extremely unusual information engineering in the annals of science was Dr. Robert A. Kehoe who was discovered by Charles Kettering during the early turmoil in commercial production of TEL. For most of his professional life, he was simultaneously the medical director of Ethyl Corporation, the director of Kettering Laboratory and professor of physiology at the University of Cincinnati. In referring to the importance of Kehoe's work, Nickerson (1954, p. 568) observed that "the antiknock compound industry might not even exist today without the work they (Kehoe and his staff) have done and are still doing." According to Robert (1983, p. 294), "Kehoe had the fate of the company (Ethyl Corporation) in his hands; if he had wavered the company would have been faced with disaster."

Robert Kehoe was an ingenious but devious man. He successfully introduced the principle of dose-response relationship to reduce the occupational exposure in the TEL manufacturing facilities. He maintained that the hazard of TEL poisoning could be controlled to an acceptable degree by reducing exposure to a safe threshold and that it was not necessary to eliminate exposure completely to keep industrial workers safe. After this initial success, he quickly established himself as the world's foremost authority and influential scientist on occupational and environmental lead hazards and wore this mantle proudly for over four decades.

For many years, Robert Kehoe, the Kettering Laboratory, and corporate America successfully established a hegemony on scientific knowledge about lead and lead poisoning in America. Clinical and scientific evidence were carefully gathered to perpetrate the Kehoe-Kettering perspective in the scientific literature and to manipulate the public's perception of the risk of automotive lead. Using professional ties in the medical establishment and the persuasive influence of lucrative grants, a network was established which successfully excluded from the scientific literature and professional reports any negative opinion pertaining to the hazards of lead to the general public. The views on lead poisoning as a public health hazard provided by the committees of the American Medical Association, the American Public Health Association, the US Public Health Service, American Association of Industrial Physicians and Surgeons, the

National Research Council and other scientific bodies were strongly influenced by the corporate interest.

Kehoe and his cohorts staunchly maintained that industries could not be responsible for lead in the body since small amounts of this element occurred naturally in the human tissues and body fluids. They claimed that the bulk of the lead was derived via the natural food chain rather than from anthropogenic sources and as such the body burden of lead in the general population had not changed in decades. Most of the lead ingested was rapidly excreted and the little that remained in the body was stored harmlessly in the bones. Increased exposure to lead only slightly increased the body burden but the increase was never likely to reach a toxicity threshold. Body defense and adaptive mechanisms (see Figure 2) made it highly unlikely that environmental exposure could induce chronic lead poisoning in the general population. Most cases of lead poisoning arose from improper nutrition which fostered the uptake and hampered the excretion of lead. Armed with a huge armamentarium of their own medical and scientific research results in support of these views, they were able to dismiss any reports or claims of chronic lead poisoning from environmental contamination as being a myth rather than a fact (Haley, 1966).

Graebner (1985, p. 155) has noted perceptibly that during this period, "the lead industries engineered the development, dissemination and perception of knowledge concerning the lead hazard. Some of the mechanisms of information engineering, such as the Lead Industry Association, frankly and openly existed to present the industry's perspective. Others, including the Kettering Laboratory, translated the industry's needs into the language of science. And still others, semi-private organizations such as the American Public Health Association and the American Medical Association, digested that science and attested to its worthiness." The suppression of pluralism in the scientific community and the total industrial domination of research into the risk of lead to the public provided the climate for unrestrained growth in production of leaded gasoline. From about 1930 onwards, the consumption of lead in gasoline increased steadily and peaked at about 270,000 tonnes in the early 1970's (Figure 3) when leaded gasoline accounted for nearly 90% of all the automobile fuel sold. It is estimated that over 7 million tonnes of lead were burned as additives in the United States between 1926 and 1985. By the 1960's when automobile horsepower reached its peak, compression ratios of 10:1 were attained, the octanes reached 100, and the lead content of premium leaded gasoline exceeded 0.78 g/L (Nriagu, 1990). Leaded gasoline remained the main force driving the automobile industry.

CLAIR PATTERSON: CHALLENGE OF THE PREVAILING SCIENCE ON EFFECTS OF AUTOMOTIVE LEAD

"Dr. Kehoe, on April 30, 1956, in temperate and scholarly fashion addressed the California Club in Los Angeles, reassuring his audience in the matter of contamination of the air by lead compounds. A key figure in the critical approach was Dr. Clair Patterson, who had settled at the California Institute of Technology as a research fellow in 1952. His earliest distinction was association with colleagues at the University of Chicago in developing methods for dating common rock minerals and determining the sequence of geological events" (Robert, 1983, p. 293; a commissioned history of Ethyl Corporation).

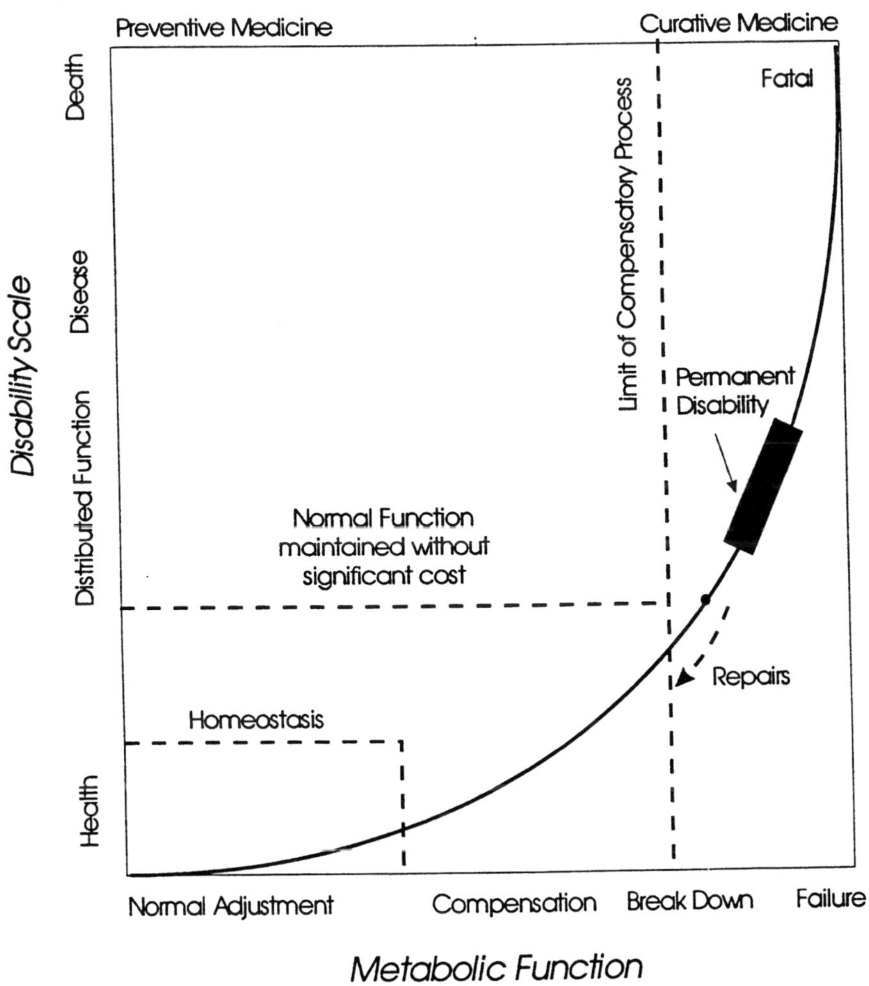

Figure 2. Concept of adaptive mechanism, impairment, and toxicity of lead (after Mastromatteo, 1984) which was perpetrated by corporate physicians for many decades. Dr. Patterson called the scheme a farce.

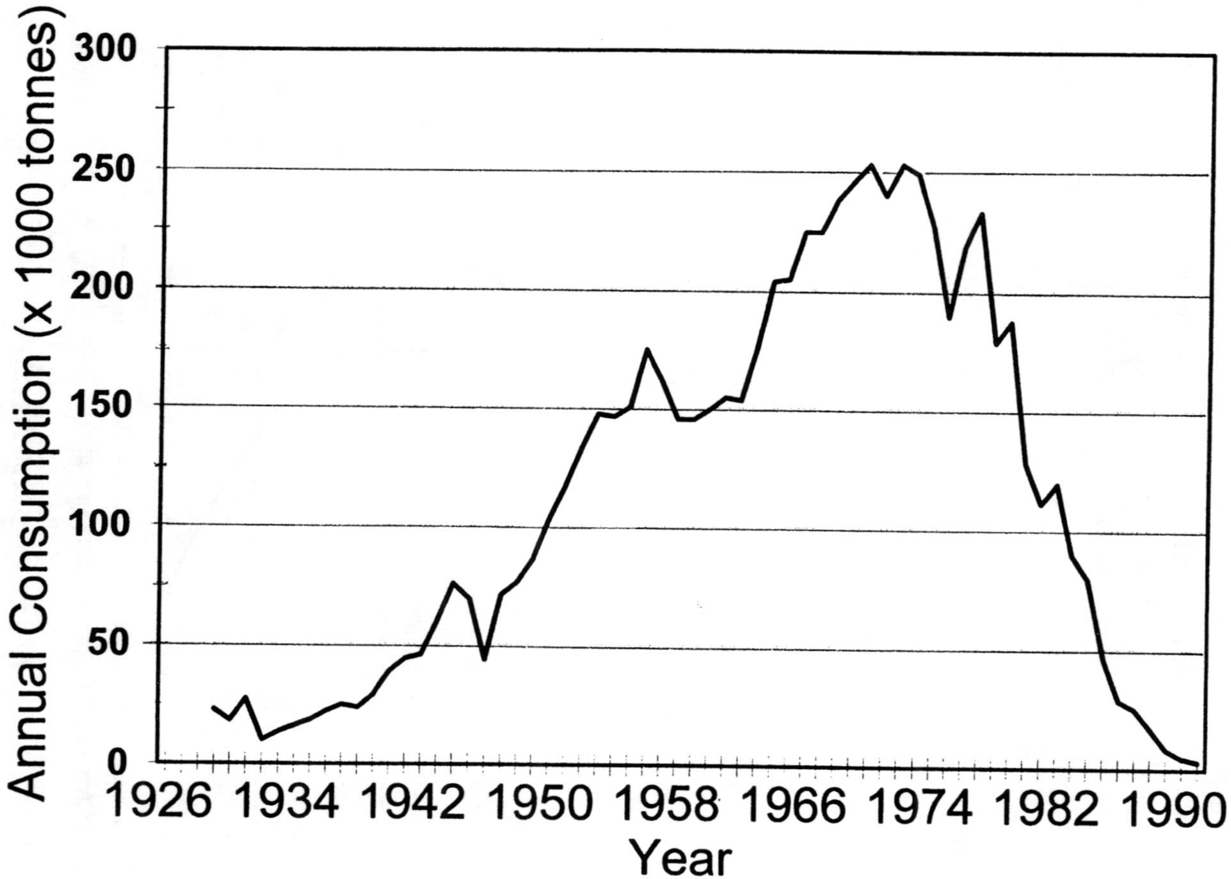

Figure 3. Historical trends in the consumption of lead in gasoline in the United States.

The excerpt above is an admission even by the Ethyl Corporation itself that Dr. Patterson played a role in the removal of lead from gasoline. Patterson's wide-ranging interests and pioneering work on the biogeochemistry of lead in the environment are discussed in detail in other chapters of this book. It suffices to note here that his science of environmental lead pollution was grounded on meticulous measurements and irrefutable methodological argument. For example, he showed that most of the published data on lead in the environment had been compromised by severe sample contamination. He pioneered the development of ultraclean laboratory procedures for determining low levels of lead in various environmental media. Using the new methods, Patterson and his colleagues were able to relate the sharp increase in the lead concentration in recent ice layers of Greenland to the deposition of industrial lead (Murozumi et al., 1969). This paper was the first to document that lead pollution was a global phenomenon.

Patterson's electrifying paper "Contaminated and Natural Lead Environments of Man" changed the focus of the lead debate and brought the dangers of environmental lead exposure to the public consciousness. He charged that the industrial use of lead is so massive that "the atmosphere of the northern hemisphere contains about 1000 times more than the natural amounts of lead" (Patterson, 1965, p. 344). He called into question the prevailing belief about blood lead levels, noting that "existing blood lead concentrations have for decades been regarded as natural, although it is well known that the average value lies only slightly below the threshold levels for classical lead poisoning" (p. 358). He continued by noting that "The threshold for damage concept, as it is applied to lead, is an ill-defined opinion unsupported by any evidence" (p. 357). His observation on possible neurotoxic effects of lead was particularly astute:

> "The basic danger involved in lead pollution is not simply that, in common with some other kinds of industrial filth, it may bring agony into our existence and shorten our lives. The course of human events is determined by the activities of the mind. Intellectual irritability and dysfunction are associated with classical lead poisoning, and it is possible, and in my opinion probable, that similar impairments on a lesser but still significant scale might occur in persons subjected to severe chronic lead insult.... (such effects) are induced by exposures to atmospheric lead concentrations corresponding to those exposures now experienced by dwellers in most large American cities" (p. 352).

In the same article, Patterson predicted that "it would be tragic, if many decades from now, it were recognized from accumulated evidence that large segments of the populations in ours and other nations had suffered needless disability and torment because early warning signs like those recognized in this report went unheeded" (p. 357). His prediction was realized much sooner than he had expected.

Patterson's (1965) paper created a panic and elicited furious reaction among the lead poisoning research establishment. Over a dozen "letters to the editor" were subsequently published essentially to excoriate and denigrate the author. Robert Kehoe (1965) observed that "Dr. Patterson has ventured, without caution, humility, or appropriate critique into what is clearly for him an alien field of biology" (p. 736) and then went on to affirm that "there is no present threat to the health of the population of the United States from the quantities of lead which occur in the food and beverages and in the ambient atmosphere or air of our communities" (p. 739). The discomfiture and anger of the lead research establishment is epitomized by the following irate letter by Dr. Herbert E. Stockinger (then Chief of Toxicology Section, Public Health Service, Department of Health, Education and Welfare, Cincinnati):

> "Dear Katharine [Boucot]. No doubt you have heard by now plenty of repercussions from the publication of Patterson's article on 'Contaminated and Natural Lead Environments of Man.' If you haven't, I have.
>
> "I had hardly returned from Europe when I received calls from industrial hygienists interested in the lead problem asking whether I was out of my head to let an article like that be published. Even while in Europe, individuals embarrassed me in discussions of this article. What I can't understand is, how did such an article not come to my attention, since it is directly in my major field of interest. I certainly would never have sanctioned its publication because all of its conclusions which are rabble rousing and are based on a single basic fallacy that because environmental lead has recently increased 30-fold than inferred natural rates se ipso facto, the American public falls for it, to quote his words, 'Residents of the U.S. today are undergoing severe classic lead insult.' This is one example of use of one false assumption one can arrive at false conclusions on health effects. There is no need to make such assumptions, as data leading to contrary conclusions are before us. (Just two weeks ago at the World Health Organization we reviewed this very problem and arrived at the conclusion based on data from 15 contributing nations, that environmental lead contribution to body burden has not altered in any significant way either blood or urinary lead contents over the past two decades).
>
> "On the other hand, this article has furnished the requisite fire to start off groups interested in public health on a large and necessary program of investigation of the matter. Is Patterson trying to be second Rachel Carson? Let's hope that this article will prove to be the first and last on Science Fiction" (Herbert Stockinger, Committee on Public Works, 1966, p. 319).

One year later, Dr. Patterson expanded his challenge during the hearings of the US Senate Subcommittee on Air and Water Pollution which was chaired by then Senator Edmund Muskie. A brief background to this meeting is pertinent here. In 1958, Ethyl Corporation requested the advice of the Public Health Service on increasing the maximum amount of tetraethyl lead from 3 to 4 ml per gallon in order to keep pace with the fuel requirements dictated by the more powerful engines. Early in 1959, the Surgeon General appointed an ad hoc committee to evaluate the proposed increase in gasoline lead in terms of its possible effects on health. The ad hoc committee, composed of Public Health officials, officials of state and local health and air pollution control agencies, and industrial authorities on hygiene, advised the Surgeon General that the proposed increase in lead apparently would pose no health hazard, but that more research was needed on levels of atmospheric lead in urban areas and on their relationship to the amount of lead in human beings. Acting on the committee's recommendation, the Surgeon General invited the automobile industry, gasoline, and lead antiknock producers to join with the Public Health Service in a survey of atmospheric lead in selected cities and of lead concentrations in blood and urine of persons living in the same cities. The Tri-City study, conducted between May 1961 and June 1962, found the average air concentrations of lead to be 1.4 $\mu g/m^3$ in Cincinnati, 1.6 $\mu g/m^3$ in Philadelphia and 2.5 $\mu g/m^3$ in Los Angeles with much higher concentrations of 14 $\mu g/m^3$ recorded in congested traffic in Cincinnati and 25 $\mu g/m^3$ on the Los Angeles freeways (Cholak, 1964). Eleven persons out of the 2,300 individuals surveyed had blood lead levels of over 60 μg/dl, then considered to be the safe threshold. The survey also found that the levels of lead in blood tended to increase as the place of residence varied from rural to central urban areas and that there appeared to be an orderly progression in blood lead levels according to the most likely concentrations of lead in the atmosphere to which the individuals were exposed. The Muskie hearings were to focus on these findings.

In his testimony, Dr. Patterson served as a forceful vent for the growing apprehension within the scientific community about corporate forces that were forestalling meaningful research on the health effects of environmental lead pollution. He charged that the source of the misinformation was an unfortunate alliance between industry, the public health departments and academics in occupational medicine and environmental health engineering:

> "It is clear, from the history of development of lead pollution problem in the United States that responsible and regulatory persons and organizations concerned in this matter have failed to distinguish between a scientific activity and the utilization of observations for a material purpose ... It is not just a mistake for public health agencies to cooperate and collaborate with industries in investigating and deciding whether public health is endangered -- it a direct abrogation and violation of the duties and responsibilities of those public health organizations" (Clair Patterson, Committee on Public Works, 1966, p. 315).

He called the Tri-City study by the consortium which "defends and promotes ideas that may be dangerous to the health of all Americans" a whitewash and noted that the reported ambient lead concentrations are "dangerous for people, especially children who live near freeways." He took particular umbrage at the categorical denial by Robert Kehoe that there had been little, if any, change in the airborne concentrations of lead in urban areas of America from 1930 to 1962.

In order to deal with what he regarded as a menace to public health, Patterson (1965, p. 358) called for "defining natural and toxic lead levels with greater care than in the past; investigating deleterious effects of severe chronic lead insult; investigating the dispersion of industrial lead into food chains; elimination of some of the most serious sources of lead pollution such as lead alkyls, insecticides, food can solder, water service pipes, kitchenware glazes, and paints; and a reevaluation by persons in position of responsibility in the field of public health of their role in this matter." Most of his suggestions have since been acted upon.

Dr. Patterson's pioneering work is now widely regarded as a key component in the demolition of the hegemonic control on research and knowledge about the hazards of lead to the general public (Corn, 1975; Graebner, 1985; Wilson, 1983; Hays, 1992). He played a major role in fostering new lines of research and in changing our perception of chronic poisoning by lead in the environment. The ultimate impact of his action on automotive lead pollution is best described in the memoirs of Dr. Harriet L. Hardy, a renowned occupational and environmental health specialist in the United States:

> "The meeting [to calm the industry down following the publication of Patterson's paper] was held in a huge room in Washington. It was a no 'shirt-sleeve' session, for there were two tables of press and more than 100 senior officials of the lead industry.... The discussion was noisy, angry, and sometimes incoherent because of emotion. After lunch there were more talks by government staff trying to make the atmosphere less tense, and the meeting broke up. A few of us were asked to stay for the press conference, a very cold-blooded affair. I almost felt sorry for one of my industrially hooked, very senior colleagues. The press asked him what his salary was and who paid it and what money supported his laboratory. Unhappily, all his funds came from one large industry. This meeting was, I think, a small-stage warning of the restrictions to come in control of environmental lead pollution in the United States. Patterson had won" (Hardy, 1983, p. 121).

The annals of history show that tetraethyl lead additive has gone from being the "gift of God" to being a curse of the gods, and that leaded gasoline became one of the largest environmentally unfriendly products to be forced out of the marketplace (mostly in the developed countries at this time). As Harriet L. Hardy said, Clair Patterson had indeed won.

REFERENCES

Cholak, J., Further investigations of atmospheric contamination of lead. *Archives of Environmental Health*, Vol. 8, pp. 314-324 (1964).

Committee on Public Works, *Hearings before a Subcommittee on Air and Water Pollution of the Committee on Public Works*, United States Senate, 89th Congress, US Government Printing Office, Washington, D.C., pp. 113-344 (1966).

Corn, J.K., Historical perspective to a current controversy on clinical spectrum of plumbism. Milbank Mem. *Fund Quarterly*, Vol. 53, pp. 93-114 (1975).

Graebner, W., Hegemony through science: information engineering and lead toxicology, 1925-1965. In: *Dying for Work*, R. Rosner and G. Markowitz (Eds.), Indiana University Press, Bloomington, Indiana, pp. 141-159 (1985).

Graebner, W., Private power, private knowledge, and public health: science, engineering and lead poisoning, 1900-1970. In: *The Health and Safety of Workers*, R. Bayer (Ed.), Oxford University Press, New York, pp. 15-71 (1988).

Haley, T.J., Chronic lead intoxication from environmental contamination: myth or fact? *Archives of Environmental Health*, Vol. 12, pp. 781-785 (1966).

Hamilton, A., Reznikoff, P. and Burnham, G.M., Tetraethyl lead. *Journal of the American Medical Association*, Vol. 84, pp. 1481-1486 (1925).

Hardy, H.L., *Challenging Man-Made Diseases*, Praeger Press, New York (1983).

Hays, S.P., The role of values in science and policy: the case of lead. In: *Human Lead Exposure*, H.L. Needleman (Ed.), CRC Press, Boca Raton, Florida, pp. 267-281 (1972).

Kehoe, R.A., Letter to the Editor concerning "Contaminated and natural lead environments in man." *Archives of Environmental Health*, Vol. 11, pp. 736-739 (1965).

Mastromatteo, E., On the concept of threshold. *Annals Am. Conf. Indust. Hyg.*, Vol. 9, pp. 331-339 (1984).

Murozumi, M., Chow, T.J. and Patterson, C., Chemical concentrations of pollutant lead aerosols, terrestrial dusts and sea salts in Greenland and Antarctic snow strata. *Geochimica et Cosmochimica Acta*, Vol. 33, pp. 1247-1294 (1969).

Nickerson, S.P., Tetraethyl lead: a product of American research. *J. Chem. Educ.*, November issue, pp. 560-571 (1954).

Nriagu, J.O., Rise and fall of leaded gasoline. *Science of the Total Environment*, Vol. 92, pp. 12-28 (1990).

Patterson, C.C., Contaminated and natural lead environments of man. *Archives of Environmental Health*, Vol. 11, pp. 344-360 (1965).

Robert, J.C., Ethyl: *A History of the Corporation and the People Who Made It.* University of Virginia Press, Charlottesville, Virginia (1983).

Rosner, D. and Markowitz, G., A "gift of God"?: The public health controversy over leaded gasoline during the 1920's. *American Journal of Public Health*, Vol. 75, pp. 344-352 (1985).

USPHS, Proceedings of a Conference to Determine Whether or Not There is a Public Health Question in the Manufacture, Distribution or Use of Tetraethyl Lead Gasoline. *Public Health Bulletin No. 158*, U.S. Public Health Service, Washington, D.C. (1925).

Wilson, D., *The Lead Scandal*. Heinemann Educational Books, London (1983).

CHAPTER 8

A CONVULSIVE CLEAR CONSCIOUSNESS: CLAIR PATTERSON AND HUMAN LEAD EXPOSURE

HERBERT L. NEEDLEMAN
University of Pittsburgh School of Medicine
Bellefield Tower
100 N. Bellefield
Pittsburgh, PA 15213

In his novel, "The Dean's December," Saul Bellow describes a character with the interesting name of Sam Beech. Beech, a geochemist, tries to convince the narrator of the book, Alex Corde, Dean of a Chicago University, that a major part of the urban unrest in the nation is due to the effects of lead on the brain. Bellow's description of this prophetic figure is intriguing:

"These scientists were diapered babies when they went public with a cause. But Beech somehow inspired respect. There was a special seriousness about him. He was physically, constitutionally serious. His head, for a body of such length was small. His face devoid of personal vanity. Light hair, grizzled and cropped Marine style, gave him an old fashioned hayseed look. His cheeks were austerely creased, his glance was dry. He was indeed an eminent man of science. That was unanimous. He had authoritatively dated the age of the earth, had analyzed the rocks brought back from the moon. Corde was beginning to think that with pure scientists, when they turned their eyes from their own disciplines, there were occasionally storms of convulsive clear consciousness; they suffered attacks of confusing lucidity."

Bellow then goes on to speak about lead poisoning:

"Here was an apocalypse--yet another apocalypse to set before the public. It wouldn't be easy. The public was used to doom warnings; seasoned. Hell--it was marinated in them. And there were evils, as someone has pointed out, that have the ability to survive identification and go on forever--money, for instance or war....Nothing was harder to get hold of than the most potent, i.e., the most manifest evils. Here science itself which was designed for deeper realization, experienced a singular failure. The genius of these evils was their ability to create zones of incomprehension. It was because they were so fully apparent that you couldn't see them."

On reading his physical description, his accomplishments, and his seriousness, I immediately recognized Beech as Clair Patterson. When I asked Patterson how this part of the book came to be written, he explained that when Bellow's wife, a mathematician, took her sabbatical at Caltech, the Pattersons and the Bellows became friends. When I asked why the sobriquet "Sam Beech," was chosen, Pat smiled knowingly.

Pat is justifiably famous among physical scientists for his work dating the age of the earth, but among epidemiologists his reputation stands upon his studies of natural concentrations of lead and the contribution of anthropogenic activity to raising everyone's body lead burdens. In promulgating this message, he didn't elect an easy path. Indeed, he achieved legendary status by forcing the National Academy of Sciences (1980) to allow him to write an independent chapter of dissent in its publication "Lead in the Human Environment." The front page of that book presents this legend: "Dr. Patterson did not want to be associated with this report. His alternative perspective appears following Chapter 5, on p. 265." The phrase "His alternative perspective" is a euphemism for a strenuous quarrel with the consensus of the NAS committee and its estimates of the dangers of lead to humans. Based upon his studies, and those of others, he asserted that we all have elevated body lead burdens, and that the NAS report was a severe miscalculation of the extent of hazard. The Academy chose not to incorporate his statements into the document, but isolated them in a separate chapter. Twelve years later, what was seen by many as a bothersome polemic now has assumed the status of a prophecy.

Scientists who have worked with Patterson in either of his spheres of activity have labored under his uncompromising standards of sample collection and analysis, and his unyielding pursuit of the facts, wherever they lead. At any gathering there will be a group comparing examples of how he bullied them because he wasn't satisfied with the rigor with which they avoided contamination: how he once turned off the scanty lightbulbs down a dark steep rocky path to a sampling site and made workers stumble to their job, because the bulbs might attract insects and the insects could, by landing on the rock samples, contaminate the specimens. Not one of his colleagues is mystified by Bellow's choice of a pseudonym.

His thesis that human activity had so contaminated the biosphere with lead, that we all were intoxicated did not endear him to the lead industry. Plainly put, they tried to paint him as a crank, and to ruin his career. At the time he published his breakthrough paper on anthropogenic sources of lead (Patterson, 1965), his geochemical studies had been the recipient of long term financial support from the oil industry. This promptly dried up, and the chemical industry petitioned Caltech, questioning his suitability as a member of the faculty (Patterson, personal communication). These attacks never affected his opinions, or bridled his expression one whit. It is some satisfaction that 30 years after the publication of his controversial paper, the bulk of scientific opinion has accepted his statements, and adjusted its position to conform to his world view about lead's ubiquity and threat. The most recent National Academy of Science's Report on Lead contains no minority chapter by Patterson. Instead, the first paragraphs adopt his position:

"Adverse health effects from exposure to lead are now recognized to be among industrialized society's most important environmental health problems. In the United States, more than 6 million preschool metropolitan children and 400,000 fetuses were believed to have lead concentrations above 10 micrograms per deciliter (μg/dl) of whole blood. That concentration has been designated by the U.S. Public Health Service as the maximum permissible concentration from the standpoint of protecting the health of children and other sensitive populations. . . A blood lead concentration of 10 μg/dl is low by comparison with the concentrations that have been associated with observable toxic reactions and that used to be widely permitted in the 1970s (e.g., 50-80 μg/dl). But it is hundreds of times *higher* than estimated blood lead concentrations in prehistoric peoples. For example, studies of bone samples of North American Indians and other preindustrial populations indicate that body burdens of lead in the general population today are 300-500 times greater than preindustrial background concentrations."

Progress in lead toxicology can be accurately measured by the shift in station of Patterson's declarations from a minority report to the opening paragraph.

Patterson was not the first to write of lead's toxicity. Two thousand years ago, Dioscerides, perhaps the first neurotoxicologist, wrote that "lead makes the mind go astray," and Pliny warned of the dangers of breathing lead fumes. Benjamin Franklin in a letter to a friend, carefully described the "dangles and gripes" (wrist drop and colic) that afflicted typesetters, tinkers and painters. Charles Dickens, in "The Uncommercial Traveler," visited a rooming house near a lead mill on the Isle of Dogs outside of London. The landlady pointed out a young woman in the corner, moaning with pain, with "her brain coming out her ear." His guide explained that it was the lead mill, and everybody knew that working there meant inevitably to become sickened. The landlady's daughter joined them, with exciting news: she had just applied to be taken on at the mill. "What could she do? Better to be ulcerated and paralyzed for eighteen pence a day, while it lasted, than see your children starve," wrote Dickens.

The past decade and a half have seen massive growth in the understanding of lead's toxicity at low dose on the central nervous system, and on many other critical systems, including the kidney, bone, endocrine function and the immune system. As studies have increased the sensitivity of their outcome measurements, and the rigor of their design, effects have been found at lower and lower doses. The gap between what is known about lead and what has been done to prevent its toxic manifestations has grown. There are a number of reasons for this. Lead poisoning has been thought to be limited to poor minority children, and the mother's caretaking often blamed for the exposure. Academic medicine has not found the problem as fascinating as the mechanics of genetic disorders, or the challenge of organ transplant biology. This is seen in many centers as a humdrum problem of small intellectual challenge. Because legislation banning the use of lead in household paint has been on the books for 20 years, and because lead has been removed from gasoline for about seven years, many believe that the problem no longer exists.

Perhaps the longest standing reason, and the most reprehensible, has been the activities by those who profit from the use of the metal to camouflage its dangers. In this effort the services of individuals from the academic world have been employed to coat the attacks with a patina of legitimacy. This is one area in which Patterson and I have shared similar experiences. In discussing some of mine, which are not unique to workers in lead toxicity, I hope to increase understanding of the reasons for the halting and glacially slow response to lead and other environmental threats. I also hope to enlarge appreciation of the tactics applied by some industries to intimidate those who would investigate their products, and to penalize scientists who penetrate these fields.

Childhood lead poisoning was first described in 1894 in Brisbane, Australia (Turner, 1897). It was believed that there were only two outcomes from the acute disease: death or complete recovery. Most pediatricians held that if a child did not die of the acute illness, he or she recovered without any lasting residua. In the 1940's, a Boston pediatrician, Randolph Byers, was referred some children with aggressive behavior, such as attacking a teacher. He recognized some as former patients of his that he had treated for lead intoxication. This drew him to follow up 20 children who had recovered from lead poisoning. He found that 19 of 20 had behavior or learning problems. The modern period of lead toxicity can be dated from this report (Byers and Lord, 1943). It is pertinent that the lead industry through their trade organization, the Lead Industries Association, approached Byers in an effort to get him to muffle his message. This failed, and they threatened him with a million dollar lawsuit.

In the mid 1960's, the role of a pediatrician confronted with a child with an elevated blood lead was relatively simple: if the level was greater than 60 μg/dl, the child was treated with a chelating agent; if the blood lead was 59, the patient was followed. The notion that there was a clear demarcation between toxicity and safety was challenged by the writings of Patterson and Byers, and when studies of American cities showed a disturbingly high prevalence of children with blood lead levels above 40 μg/dl, the question of unrecognized lead toxicity began to be raised by public health authorities.

Research into the damage from lead at silent doses slowly began to proceed. In the 1970's, a few investigators reported that children with elevated blood lead levels, but no symptoms, had lower IQ scores (Perino and Ernhart, 1974; de la Burde and Choate, 1975). A few reports of no effect of lead at low dose were also published (Landsdown et al., 1986; Ernhart et al., 1981). Most of these reports were vexed by small sample size, poor measures of exposure and outcome, and inadequate control of confounding. My colleagues and I, recognizing these design difficulties, devised a study that attempted to confront these problems. I had shown that the tooth was a better marker of past lead exposure than the customary measure, blood lead concentration (Needleman et al., 1974). Using this strategy, we collected teeth from about 2500 primary school children in two towns near Boston, Massachusetts, and examined the psychological performance of children with the highest and lowest lead levels. We showed that after controlling for parental education and IQ, family size, and parental occupation, children with high dentine lead levels had lower IQ scores, poorer attention, and impairment on a number of other psychological performances. When we compared the ratings of teachers on eleven classroom behaviors, we found that as dentine lead increased, poor performance increased on each measure (Figure 1). The data were influential in a number of regulatory decisions, including the Air Lead Standard, and the decision to remove lead from gasoline.

In the 1980's, studies from around the world began to show lead effects at lower and lower doses (Fulton et al., 1987; Hatzakis et al., 1987; Fergusson et al., 1988; Hansen et al., 1989). The lead industry asserted that lead was not the cause of deficit; impaired children, they argued, exposed themselves to more lead. To evaluate this, studies of children forward from birth were conducted. These also showed that lead was neurotoxic (Bellinger et al., 1987; Dietrich et al., 1987; Baghurst et al., 1992). This flood of new information caused the federal government to revise its definition of lead poisoning and its response levels. Lead toxicity in childhood had been defined as a blood lead level greater than 60 μg/dl in the 1960's, but in the ensuing 30 years it would drop to 10 m μg/dl (Centers for Disease Control, 1991a).

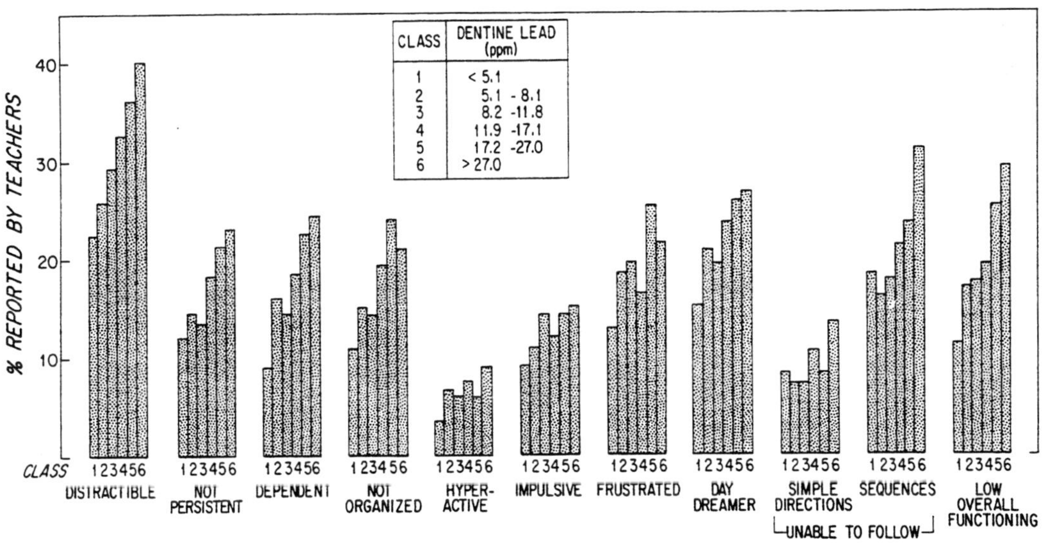

Figure 1. The rate of teachers' reports of negative classroom behaviors in children classified by dentine lead level. The Instrument was an eleven item forced-choice questionnaire. The number of students was 2146. Teachers were blind to the students' lead levels.

The Agency for Toxic Substances, given the charge to measure the sources of lead and the number of exposed Americans, estimated that almost 17% of children under 5 years had blood lead levels greater than 15 μg/dl. The risk was increased both by poverty and minority status. The official estimate of the rate of elevated blood lead levels in African-American children in poverty was 55% (Agency for Toxic Substances and Disease Control, 1988).

Pressure began to build in the regulatory agencies and in congress to attack this massive problem. The CDC took a revolutionary step: it moved from case finding to true primary prevention with the publication of the document "A Strategic Plan to Eliminate Childhood Lead Poisoning" (Centers for Disease Control, 1991b). This represented a fundamental change in federal policy, and a new chapter in the lead story.

At each step in the regulatory process, the lead industry attempted to impede the control of lead and to cloud the scientific interpretation of the evidence. To this end, it funded research to show that lead was an essential trace element (Reichlmayr-Lais and Kirchgessner, 1981), and criticized the growing body of data from epidemiological studies of children, and experimental studies of animals. When reviewing epidemiological studies, they criticized the inability to control for all covariates, known and unspecified. When presented with strong data on lead's effects in animal studies where covariate control was not an issue, the automatic response of industry was that extrapolation from animal studies to man was unacceptable to make the case. The pairing of these two criticisms combines to erect an impregnable wall of pseudoscientific agnosticism. Multivariate space is, after all, infinite and research budgets are not. The number of subjects in a study must be less than the list of possible confounders. All epidemiological studies as a result are open to the criticism of unmeasured confounders. In general we do not give toxins to humans. The only model of man not vulnerable to the criticism of species differences is man; this criticism can be applied to any toxin. This logic, if accepted, means that we can never demonstrate that any agent is toxic.

Another tactic of the industry has been to employ academics to impeach the integrity of scientific studies and investigators. This is a tactic experienced by Patterson and me. In 1983, questions were raised about the probity of my data before the Environmental Protection Agency. As a result, a committee of scientists representing the EPA examined my primary data from the 1979 study. Their report was then reviewed by the Clean Air Science Advisory Counsel of EPA, who asked me to testify before them. At the conclusion of this meeting, CASAC filed a report which accepted the work of my group as valid and useful in determining the allowable amount of lead in the environment.

This did not stop the lead industry. Seven years later I was asked by the Department of Justice to serve as an expert witness in a suit against the owners of a lead mill and mine to recover funds to make the site safe for habitation. Appearing as witnesses for the defense were two psychologists. One, Dr. C. Ernhart, had reported in 1974 that she had found lead effects on intelligence in asymptomatic children. Later, she revised her opinion in a follow up of her subjects (Ernhart et al., 1981), and deprecated the lead effect she had found. Following this she was awarded a seven year grant from the International Lead Zinc Research Organization, the industry trade group. The other psychologist, S. Scarr, had served as a member of the EPA visiting team in 1983, but now appeared as a witness for the industry. They visited my lab and inspected the same printouts that Scarr had seen seven years earlier.

After leaving my lab, they wrote a report that accused me of manipulating my data to create a false impression that lead was toxic. The specific charges were that I selected subjects with knowledge of the outcome and lead burden to maximize the lead effect, that I ignored the effect of age on IQ scores, and that I applied a large number of tests, and only reported those that showed the effect I was seeking. Xerox copies of their screed were widely circulated. I received a copy from a plaintiff's lawyer who obtained it from the defense, the insurers of a leaded property, in a suit for damages. It became the subject of a number of newspaper columns. The charges were submitted to the National Institutes of Health, and in the winter of 1991, I was notified that the NIH Office of Scientific Integrity had directed the University of Pittsburgh to begin an inquiry into the methods I had used to conduct the study I had reported 12 years earlier. The cover letter to the NIH, suggesting that I should be investigated, was not signed by either Scarr or Ernhart, but by David Genesson, an attorney with the firm of Hunton and Williams, the law firm that represented the Ethyl Corporation.

I was puzzled both by the choice of the site, and the timeliness of the investigation. The research had been conducted when I was a member of the Harvard Medical School faculty, and was completed in 1979, before there were any regulations about scientific misconduct in place. Nevertheless, I turned over all of my data to the committee appointed by the dean of the medical school. After weeks of thorough examination of my data, the panel stated that they found no evidence of scientific misconduct. Strangely enough, they also stated that they could not "rule out scientific misconduct in terms of misrepresentation of the data." A full investigation was scheduled.

I asked the dean to provide me with a full and detailed list of specific charges, and to appoint to the hearing board experts in the disciplines of epidemiology and behavioral toxicology. I also asked him to replace two members of the board on the grounds of potential conflicts of interest. One board nominee had served on many committees with Dr. Scarr, and frequently cited her work in his, and the other had been chairman of Environmental Health at the same institution as Dr. Ernhart. The dean denied all of my requests. I was given no list of specific charges; I was forced to infer what I was accused of. The faculty handbook at the university states that in scientific misconduct hearings, the respondent is given the opportunity to question his accusers. I asked that these hearings, conventionally closed affairs, be held in the open, so that the scientific community, and my colleagues on the faculty could weigh the evidence for themselves. I sought and received the unanimous support of the Faculty Assembly and Senate in this demand. Nevertheless, the administration maintained that the hearings would be closed. This was necessary, they claimed, "to protect the process." I argued that the process lacked a nervous system, and that it was the university's obligation to protect people. Only after a strong resolution by the Faculty Senate, and after I filed a suit in the federal court, did the university consult with the NIH and agree to open the hearings.

My accusers suddenly became reluctant to attend. This led to prolonged negotiations between them and the university administration. The hearings were held in April of 1991. I confronted the three allegations (prior selection of subjects, lack of control for age, and multiple comparisons) in the document circulated by my accusers.

To evaluate their charge that I knew the outcome of my subjects and their lead level, and selected subjects with this knowledge, I showed them a piece of computer code from the printouts they had examined. It said:

SELECT IF PB EQ 1 OR PB EQ 2 AND PBCONSD EQ2 AND PLUMBISM EQ2 AND CHELATED EQ2LANGHME EQ1 AND HMETGHR EQ1

This means: "Select if lead level equal high or low, and head injury equal 'no', and plumbism equal 'no' and English is the first and only language in the home and the subject was discharged with the mother." This proved conclusively that the subjects were excluded on criteria that were identified before the study was begun and could not have been influenced by me or any other member of my research team. This piece of code was on the first page of the printouts, and was in the header of each data run. It appeared 24 times in the printouts that my accusers examined.

I asked Dr. Ernhart:

DR. NEEDLEMAN:...does this look familiar to you at all?

DR. ERNHART: I don't recall having seen this (Continuation of Public Hearing, 1992)

I next turned to the issue of control of confounders. In the calculation of the Wechsler IQ score for children, the instrument we employed, the age of the child is entered and adjusted for.

I asked Dr. Ernhart if this was not the case. She said that the adjustment was not always adequate and age should be considered. I then asked her if she had entered age into the models in her papers on lead and IQ. She refused to answer, saying the question was irrelevant. She did not adjust for age in her studies.

The final allegation, that I applied multiple tests and only reported the positive ones, is demonstrably untrue on examination of the paper. It was not mentioned by my accusers at the hearing, and was not considered by the hearing board.

The final report of the hearing board stated:

"The Hearing Board finds that there was no fabrication on the part of Dr. Needleman in the conduct of his research. What is particularly important is that we found no evidence of deliberate falsification of data, procedures or analysis that would bias the results in favor of finding a harmful effect for lead on the neuropsychological development of young children.

The Board also finds that the data file which Dr. Needleman has shared with us and others contains the primary data available for the 270 children for whom psychological data were collected and that it accurately reflects the data actually collected between 1975 and 1978. The Board is also confident that these data reveal a harmful effect of lead among the 270 children with a variety of analytical approaches.

The Board found that most aspects of this study that were being disputed by the complainants and Dr. Needleman's other critics were honest differences in how such studies should be conducted or in interpretation and judgments of data."

In the body of the report, this statement is found:

"Finally, perhaps the most remarkable feature of the printout is that it exists in the first place. If Dr. Needleman had the intent of fabrication and falsification (of results), much of the printout would have been buried long ago. Indirectly, the Board sees this as inconsistent with the *intent* to be scientifically fraudulent."

In the most recent National Academy of Sciences report on lead, a reanalysis of my data was evaluated. The report concluded:

"Schwartz [EPA scientist who reanalyzed my data] also examined the association between dentin lead and other outcomes. These outcomes might be expected to covary with IQ and association with them further indicates that the IQ results were not anomalous in these data. In their models that control for all the covariates used in the IQ regressions, dentin lead was associated with Teacher Rating Score, with Piaget Mathematics, and Reading scores, with Seashore Rhythm, and other neurobehavioral outcomes. These findings suggest that the association between dentin lead and intellectual development in the data of Needleman and co-workers is strong and robust."

In his important book "The Structure of Scientific Revolutions," Thomas Kuhn (1962) depicts the conduct of science as something other than a smooth, self-correcting march toward some single-valued portrait of reality. It is a human endeavor in which contesting world views struggle for supremacy, in which the very language, tools, and acceptable areas of investigation are governed by the prevailing conceptual consensus. When the profit motive is entered into this mix, the untidy set of proceedings is severely perturbed, and not in a random manner. We have seen how investigators who challenge the safety of a widely used commercial product can have their investigations challenged, can have attempts made to damage their careers and reputations, and how the process of controlling scientific misconduct itself can be bent to harass these investigators. If this can be attempted with established investigators, what are the effects upon less secure junior scientists who are thinking of studying the toxic effects of some commercial product?

Patterson understood these hazards; yet they deterred him not at all. He paid some penalty for his storms of convulsive clear consciousness. Blessedly, in the last years of his life, he was able to see his role in identifying the true dimensions of lead toxicity recognized and validated. His colleagues and students know what he has meant to two separate and important fields of science. It remains to liquidate a portion of our debt to him by extinguishing this threat of lead to the brains and futures of our children.

References

Agency for Toxic Substances and Disease Control, *The nature and extent of lead poisoning in children in the United States: a report to Congress*. Dept. of Health and Human Services, Atlanta, Georgia (1988).

Baghurst, P.A., McMichael, A.J., Wigg, W., Vimpani, G.V., Robertson, E.F., Roberts, R.J. and Tong, S.I., Environmental exposure to lead and children's intelligence at the age of seven years. *New England Journal of Medicine*, Vol. 327, pp. 1280-1284 (1992).

Bellinger, D., Leviton, A., Waternaux, C., Needleman, H.L. and Rabinowitz, M., Longitudinal analyses of prenatal and postnatal lead exposure and early cognitive development. *New England Journal of Medicine*, Vol. 316, pp. 1037-1043 (1987).

Byers, R.K. and Lord, E.E., Late effects of lead poisoning on mental development. *American Journal of Diseases in Children*, Vol. 66, pp. 471-483 (1943).

Centers for Disease Control, *Preventing Lead Poisoning in Young Children.* USDHSS/PHS/CDC (1991a).

Centers for Disease Control, *Strategic Plan for the Elimination of Childhood Lead Poisoning*. Dept. of Health and Human Services, Atlanta, Georgia (1991b).

Continuation of the Public Hearing of Scientific Misconduct Allegations against Herbert Needleman, M.D., April 13, 1992.

de la Burde, B. and Choate, M.S., Early asymptomatic lead exposure and development at school age. *Journal of Pediatrics*, Vol. 87, pp. 638-642 (1975).

Dietrich, K.N., Krafft, K.M., Bornschein, R.L., Hammond, P.B., Berger, O. et al., Low-level fetal lead exposure effect on neurobehavioral development in early infancy. *Pediatrics*, Vol. 80, pp. 721-730 (1987).

Ernhart, C.B., Landa, B. and Schell, N.B., Subclinical levels of lead and developmental deficit--A multivariate follow-up reassessment. *Pediatrics*, Vol. 67, pp. 911-919 (1981).

Fergusson, D.M., Fergusson, J.E., Horwood, L.J. and Kinzett, A longitudinal study of dentine lead levels, intelligence, school performance and behavoir. Part II. Dentine lead and cognitive ability. *Journal of Child Psychology and Psychiatry*, Vol. 29, pp. 793-809 (1988).

Fulton, M., Raab, G., Thomson, G., Laxen, D., Hunter, R. and Hepburn, W., Influence of blood lead on the ability and attainment of children in Edinburgh. *Lancet*, Vol. 1, pp. 1221-1226 (1987).

Hansen, O.N., Trillingsgaard, A., Beese, I., Lyngbye, T. and Grandjean, P., A neuropsychological study of children with elevated dentine lead level: assessment of the effect of lead in different socioeconomic groups. *Neurotoxicology and Teratology*, Vol. 11, pp. 205-213 (1989).

Hatzakis, A., Kokkevi, A., Katsouyanni, K. et al., Psychometric intelligence and attentional performance deficits in lead-exposed children. In: *International Conference on Heavy Metals in the Environment*, CEP Consultants Ltd., Edinburgh, Scotland, pp. 204-209 (1987).

Kuhn, T., *The Structure of Scientific Revolutions*. University of Chicago Press, Chicago, Illinois (1962).

Lansdown, R., Yule, W., Urbanowicz, M. and Hunter, J., The relationship between blood lead concentrations, intelligence, attainment and behaviour in a school population: The second London study. *International Archives of Occupational and Environmental Health*, Vol. 57, pp. 225-235 (1986).

National Academy of Sciences/National Research Council, *Lead in the Human Environment.* National Academy Press, Washington, D.C. (1980).

Needleman, H.L., Davidson, I., Sewell, E. and Shapiro, I.M., Subclinical lead exposure in Philadelphia schoolchildren: Identification by dentine lead analysis. *New England Journal of Medicine*, Vol. 205, pp. 245-248 (1974).

Patterson, C.C., *Personal communication.*

Patterson, C.C., Contaminated and natural lead environments of man. *Archives of Environmental Health*, Vol. 11, pp. 344-360 (1965).

Perino, J. and Ernhart, C.B., The relation of subclinical lead level to cognitive and sensorimotor impairment in black preschoolers. *Journal of Learning Disabilities*, Vol. 7, pp. 26-30 (1974).

Reichlmayr-Lais, A.M. and Kirchgessner, M., Hematological changes with alimentary lead deficiency. *Annals of Nutrition and Metabolism*, Vol. 25, pp. 281-286 (1981).

Turner, A.J., Lead poisoning among Queensland children. Australian Medical Gazette, Vol. 16, pp. 475-479 (1897).

CHAPTER 9

ANALYZING POLAR ICE AND SNOW FOR HEAVY METALS AND OTHER TRACE ELEMENTS: TWO DECADES OF FRUITFUL INTERACTION WITH CLAIR PATTERSON

CLAUDE F. BOUTRON
Laboratoire de Glaciologie et Géophysique de l'Environnement du CNRS
54, Rue Molière, Domaine Universitaire, B.P. 96, 38402 Saint-Martin d'Hères, France
and, UFR de Mécanique, Université Joseph Fourier de Grenoble
Domaine Universitaire, B.P. 68,38041 Grenoble, France

1. BEGINNING TO INVESTIGATE TRACE ELEMENTS IN ANTARCTIC SNOW AND HEARING ABOUT CLAIR PATTERSON

It was in 1968, when I was a 4th year Physics student at the University of Grenoble, that I first heard about an unusual small laboratory which had been created a few years before by Prof. Louis Lliboutry in Grenoble. This laboratory was one of the very first in the world devoted to Glaciology, at that time mainly alpine Glaciology. I was immediately attracted by this rather new field of research, which looked much more stimulating to me than classical Physics, and I started to prepare for the new Geophysics and Geochemistry degree which had just been created by Prof. Lliboutry.

One of the lecturers was Claude Lorius, who was presenting to us the very limited investigations on the trace element content of Antarctic and Greenland ice and snow which were then available. I was fascinated by this field of investigation, and soon after became one of the first two graduate students of Claude Lorius who was working near Paris. To begin with, my main goal was to develop improved clean and sensitive analytical procedures to get reliable data for sea and soil derived elements such as Na, Cl, Mg, K, Ca, Mn, and Fe in surface snow samples. The samples were collected in collaboration with Soviet investigators in East Antarctica during a traverse from the coastal station Mirny to the inland station Vostok.

Both Claude Lorius and myself had very little, if any, experience in trace elements analysis, so it was decided to develop the analytical procedures in a laboratory (Service d'Etudes Analytiques) of the French Atomic Energy Commission at Fontenay aux Roses near Paris, which had experience in analytical chemistry. There, I worked under the very stimulating guidance of René Platzer and Georges Baudin, who showed great interest for this inexperienced student who wanted to quantitatively analyze trace elements at the then very unusual ng/g level.

Very soon, I heard about two already well known scientists who were real pioneers in the field: Edouard Picciotto, at the University of Brussels, and... Clair Patterson at the California Institute of Technology. Pat sent us a very impressive (about 100 pages!) preprint of the paper describing his ongoing work with Masayo Murozumi and James Chow on Greenland and Antarctic snow and ice which was published later in Geochimica et Cosmochimica Acta (Murozumi et al., 1969). I remember quite well having spent hours trying to understand this preprint. I must say that I did not fully realize at that time how crucial the very meticulous descriptions of the sampling and analytical procedures were as given in that unusually lengthy paper.

Anyway, from the information which I got both from the work of Edouard Picciotto's group (Hanappe, 1970; Vosters, 1971; Vosters et al., 1970) and from Pat, we managed to develop rather clean analytical procedures. We were among the very first in France to purchase (in 1969) specially designed laminar flow clean benches for performing analytical chemistry, not for micro-electronics. This resulted in a great breakthrough on the quality of the data for Na, Cl, Mg, K, Ca, Fe, and Mn in surface Antarctic snow (Boutron et al., 1972), as illustrated in Table 1. Our concentrations were much lower than the ones previously published by Soviet investigators, although it appeared later that for K and Ca they were still slightly artificially high as discussed in Boutron (1979). These new data made it no longer necessary to invoke unproven theories, such as that on the enrichment of sea derived aerosols for alkali and alkaline earths, to account for the observed concentrations of these species in Antarctic snow (Boutron, 1979).

Table 1. Surface snow collected between 800 and 1300 km from the sea coast along the Mirny-Vostok axis, East Antarctica: comparison between the data of Matveev (1970) and the improved data which we obtained in the early 1970's (Boutron et al., 1972).

Reference		Measured concentration (ng/g)			
		Cl	Na+K	Mg	Ca
Matveev	(1970)	340	170	120	310
Boutron et al.	(1972)	90	23	7.3	2.8

Pat even agreed to send us aliquots of some of the now famous samples which he analyzed with Masayo Murozumi and James Chow (Murozumi et al., 1969). In this way, we could check the accuracy of our procedures for Cl, Na, Mg, K, and Ca (at that time we were very far from being able to investigate Pb at the pg/g level). As shown in Table 2, our results proved to be in excellent agreement with theirs, which we were very proud of!

Table 2. Comparative analysis of Cl, Na, Mg, K, and Ca in a sample of Antarctic snow collected by Murozumi et al. (1969) at New Byrd Station, West Antarctica (sample BYR 66-A-1).

Reference	Measured concentration (ng/g)				
	Cl	Na	Mg	K	Ca
Murozumi et al. (1969)	61	31	4	1.5	1.5
Boutron et al. (1972)	62	29	3.8	1.2	1.5

In 1970, I had also the opportunity to meet Pat at Caltech for the first time during a short stay in California. Unfortunately, I was too new in the field to fully benefit from the advice which Pat gave me to improve our analytical procedures. Anecdotally, I remember quite well that Pat showed me the first large supermarket I ever saw!

2. Emulating the Work of Murozumi et al. (1969) for Heavy Metal Analysis: A Decade of Hard Work without Complete Success (1971-1981)

In 1970, I got a permanent position as Assistant Professor at the University of Grenoble, which allowed me to continue my research work on a long-term basis in the group of Claude Lorius. Since we had been rather successful for sea-derived and soil-derived elements, we decided to try to go one step further and to shift our efforts to heavy metals. This required a major change in our procedures since we were going from the ng/g level down to the pg/g level, and we knew from the pioneering work of Murozumi et al. (1969) that it would be a formidable challenge.

During the following years, I made a massive effort to emulate the work of Murozumi et al. (1969). Probably too ambitious, I tried to do it not only for Pb, but also for several additional heavy metals. A clean laboratory was designed and built in the new building of the Laboratoire de Glaciologie et Géophysique de l'Environnement at the campus of the University of Grenoble. Original procedures were developed for the production of ultrapure water, the cleaning of labware, the preconcentration of the samples, and other tasks. Numerous new samples of snow were collected using improved clean procedures at various locations in East Antarctica, West Antarctica and Greenland; among the most interesting ones were time series covering periods up to one century obtained at Dome C and at the geographic South Pole.

Considerable time was devoted to improving the cleanliness of all the procedures, from field sampling to final laboratory analysis, and also the determination of the blanks. But although considerable progress was achieved, it appeared later that I was not fully successful. A large part of the heavy metals data for Pb, Cd, Zn, Cu, and Ag which I published during that period until 1981 (see for instance Boutron and Lorius, 1979) were to a variable extent plagued by contamination problems. This is illustrated in Table 3, which compares, as an example, the Pb, Cd, and Cu data published at that time for surface snow collected at stake D 80, about 430 km inland of Dumont d'Urville station, with data obtained one decade later using fully reliable procedures. It can be seen that the oldest data were still in considerable positive error, which was a big disappointment considering the amount of work (and money) that had been spent!

Table 3. Surface Antarctic snow collected at Stake D 80, 430 km inland of Dumont d'Urville station along the coast-interior axis Dumont d'Urville-Dome C: comparison of the data obtained using procedures that were not completely reliable (Boutron, 1981) with the most recent data obtained using improved reliable procedures (Boutron and Patterson, 1987; Boutron et al., 1990, 1993).

Reference	Measured concentration (pg/g)		
	Pb	Cd	Zn
Boutron (1981)	67	12	244
Boutron and Patterson (1987)	2.3	-	-
Boutron et al. (1990)	-	0.07	-
Boutron et al. (1993)	-	-	6.6

Pat was quite angry about this French scientist from Grenoble who, during all these years, was publishing unreliable heavy metals data in various well-known journals. Soon, it became obvious that the only way to improve the situation was that I come to work directly with Pat at Caltech, as we discussed during a field party at Etna volcano in Sicily in the spring of 1980. This need for this collaboration became even more evident by the work of Amy Ng and Clair Patterson in the late 1970's (Ng and Patterson, 1981): it confirmed beyond a doubt that all heavy metals data for Antarctic and Greenland snow and ice obtained outside Pat's laboratory were erroneously high because of contamination problems.

3. Working with Clair Patterson at the California Institute of Technology (1981, 1984-1985)

Overall, I spent 18 months (6 months in 1981 and one year in 1984-1985) working with Pat at Caltech. It was a unique period of my life, which proved to be exceedingly fruitful.

Shortly before I first arrived in Pasadena in July 1981, Pat had been the first one to tackle successfully the key problem of analyzing Pb in sections of Antarctic and Greenland deep ice cores (Ng and Patterson, 1981). Such data were urgently needed to establish the natural pre-industrial concentrations of Pb in our atmosphere, thereby providing firm reference levels against which to assess modern time trends.

They were able to decontaminate the highly contaminated core sections from the Camp Century and Byrd deep ice cores using an original chiselling procedure, which allowed them to get successive veneer layers of ice in progression from the very dirty outside to the pristine center. Pat was the first to propose checking the integrity of the most central part of the core by investigating changes in Pb concentration from the outside to the center of each core section. He postulated that the value measured in the central part of the core can be considered representative of the original concentration in the ice only if there is a clear plateau of concentrations obtained in several successive central layers.

Of course, the main scene of my work with Pat was the famous clean laboratory on the third floor of the North Mudd building. I took several photographs inside the laboratory; some of them are shown at the end of this chapter.

The goal of my work was to improve the analytical procedures developed by Amy Ng and Pat, and then to use them extensively for analyzing various Antarctic snow and ice samples. These included surface snow blocks, a shallow (11 m) snow core drilled with an all plastic acid-cleaned hand-operated auger, various sections of the 905 m Dome C deep ice core, and a few sections of the 2083 m Vostok deep ice core. All of these samples had been collected during international field work of the French Polar Expeditions, the US National Science Foundation Expeditions, and the Soviet Antarctic Expeditions. The samples were first brought to France and then transported to California in a frozen state.

The chiselling procedure proved to be excellent, but it was very time consuming. It required extensive cleaning of hundreds of polyethylene sheets, of thousands of short and shoulder length polyethylene gloves, and other items. I spent hundreds of hours continuously cleaning all these items whose cleanliness was so critical.

The chiselling itself was quite an unusual business: Pat and I worked together (four hands were needed), our bodies entirely wrapped in large polyethylene bags whose outside had been cleaned in acid, as shown in Figure 1 (a, b, and c). Our hands and arms were kept inside acid-cleaned polyethylene gloves which were changed each time we went from one ice or snow layer to the next one (usually, we were chiselling 4 to 6 veneer layers before reaching the inner core). The chisels were made out of stainless steel, which at the beginning seemed to me to be a strange choice but in fact gave excellent blanks for Pb (and later also for Cd, Zn, Hg and other metals).

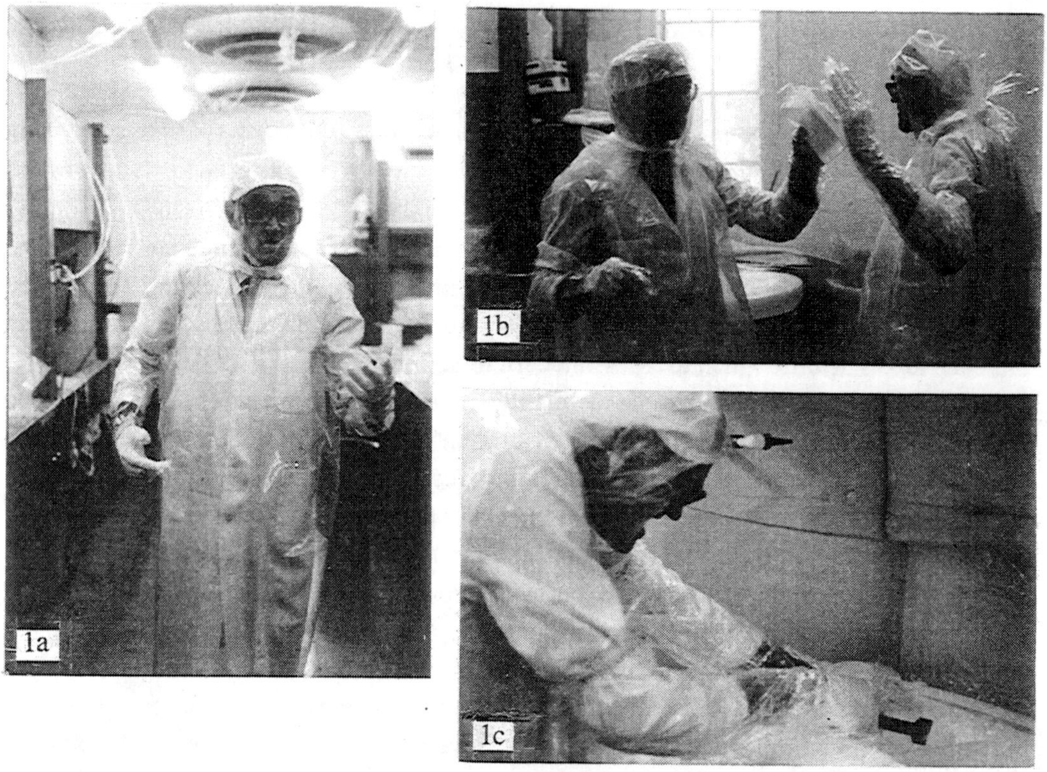

Figure 1. Mechanical de-contamination of a section (depth 500 m ; 14400 years BP) of the Dome C deep Antarctic ice core (10 June 1985). (a) Clair Patterson inside his acid-cleaned polyethylene bag, ready for action, (b) Claude Boutron washing the acid cleaned polyethylene gloves of Clair Patterson with ultrapure quartz distilled water, (c) Chiselling the core section with a stainless steel chisel. Photos by C. Boutron.

Figure 2. Clair Patterson's thermal ionization mass spectrometer (a) The machine itself (b) The control panel. Photos by C. Boutron.

The chiselling of a single core section took about 5 hours, excluding the time needed to clean the equipment. It might have been a boring occupation. In fact, it was not: during these hours together inside the clean laboratory we had the unique impression of creating new science, being the first ones to analyze Pb in such ancient ice which had been formed so many thousand years ago at an early period of human kind.

I also spent hundreds of hours downstairs running the samples with Pat's old mass spectrometer, as shown in Figures 2a and b. I greatly enjoyed running this fully manual machine, which allows you to really understand what you are doing, contrary to what happens often with modern automated equipment.

Each separate veneer layer and inner core was analyzed separately for Pb, allowing us to study in full detail changes in Pb concentration when going from the outside to the center of most core sections. Examples of such outside-inside profiles are shown in Figure 3.

Our work proved to be very successful for Pb, the main achievement being to get the first time series of Pb concentrations in Antarctic ice through the last full climatic cycle, as shown in Figure 4 (Boutron and Patterson, 1986; Boutron et al., 1987).

Furthermore, our work proved to be unexpectedly successful for the investigation of several other heavy metals such as Zn (Boutron et al., 1990), Cd (Boutron et al., 1993) and Hg (Vandal et al., 1993). This was due to the hundreds of hours I spent cleaning large numbers of polyethylene bottles of various sizes; they allowed us to systematically keep aliquots of these valuable samples which were transported frozen back to France. On the whole, these samples chiselled at Caltech have been responsible for the largest part of our present knowledge on the occurrence of heavy metals in Antarctic snow and ice, giving more than fifteen publications in peer-reviewed journals (including two in "Nature"). Figures 5-8 show the clean laboratory and equipment used in this work.

Even more important, these long hours spent working directly with Pat have allowed me to transfer to my home laboratory in Grenoble the ultraclean procedures which Pat had developed at Caltech during several decades of pioneering work. This has now opened the way to a wealth of new reliable data, including data for Greenland (Boutron et al., 1991; Rosman et al., 1993; Hong et al., 1994).

REFERENCES

Boutron, C.F., Alkali and alkaline earth enrichments in aerosols deposited in Antarctic snows. *Atmospheric Environment*, Vol. 13, pp. 914-924 (1979).

Boutron, C.F., Nouvelles données sur le contenu en métaux traces des neiges de l'Antarctique, du Groenland et de l'île de Devon. *Zeit. Gletscherk. Glazialgeol.*, Vol. 17, pp. 79-113 (1981)

Boutron, C.F. and Lorius, C.J., Trace metals in Antarctic snows since 1914. *Nature*, Vol. 277, pp. 551-554 (1979).

Boutron, C.F. and Patterson, C.C., Lead concentration changes in Antarctic ice during the Wisconsin/Holocene transition. *Nature*, Vol. 323, pp. 222-225 (1986).

Boutron, C.F. and Patterson, C.C., Relative levels of natural and anthropogenic lead in recent Antarctic snow. *Journal of Geophysical Research*, Vol. 92, pp. 8454-8464 (1987).

Boutron, C.F., Echevin, M. and Lorius, C., Chemistry of polar snows, estimation of rates of deposition in Antarctica. *Geochimica et Cosmochimica Acta*, Vol. 36, pp. 1029-1042 (1972).

Boutron, C.F., Patterson, C.C., Petrov, V.N., and Barkov, N.I., Preliminary data on changes of lead concentrations in Antarctic ice from 155,000 to 26,000 years BP. *Atmospheric Environment*, Vol. 21, pp. 1197-1202. (1987).

Boutron, C. F., Patterson, C.C., and Barkov, N.I., The occurrence of zinc in Antarctic ancient ice and recent snow. *Earth and Planetary Science Letters*, Vol. 101, pp. 248-259 (1990).

Boutron, C.F., Görlach, U., Candelone, J.P., Bolshov, M.A., and Delmas, R.J., Decrease in anthropogenic lead, cadmium and zinc in Greenland snows since the late 1960's. *Nature*, Vol. 353, pp. 153-156 (1991).

Boutron, C.F., Rudniev, S.N., Bolshov, M.A., Koloshnikov, V.G., Patterson, C.C., and Barkov, N.I., Changes in cadmium concentrations in Antarctic ice and snow during the past 155,000 years. *Earth and Planetary Science Letters*, Vol. 117, pp. 431-441 (1993).

Hanappe, F., Contribution à la chimie des neiges Antarctiques. *Estimation du taux de déposition de matière extraterrestre*, Thesis, Université Libre de Bruxelles, Belgium (1970).

Hong, S., Candelone, J.P., Patterson, C.C. and Boutron, C.F., Greenland ice evidence of hemispheric pollution for lead two millennia ago by Greek and Roman civilizations. *Science*, Vol. 265, pp. 1841-1843 (1994).

Matveev, A.A., Chemical hydrology of regions of East Antarctica. *Journal of Geophysical Research*, Vol. 75, pp. 3686-3690 (1970).

Murozumi, M., Chow, T.J. and Patterson, C.C., Chemical concentrations of pollutant lead aerosols, terrestrial dusts and sea salts in Greenland and Antarctic snow strata. *Geochimica et Cosmochimica Acta*, Vol. 33, pp. 1247-1294 (1969).

Ng, A. and Patterson, C.C., Natural concentrations of lead in ancient Arctic and Antarctic ice. *Geochimica et Cosmochimica Acta*, Vol. 45, pp. 2109-2121 (1981).

Rosman, K.J.R., Chisholm,W., Boutron, C.F., Candelone, J. P. and Görlach, U., Isotopic evidence for the source of lead in Greenland snows since the late 1960's. *Nature*, Vol. 362, pp. 333-335 (1993).

Vandal, G.M., Fitzgerald, W.F., Boutron, C.F. and Candelone, J.P., Variations in mercury deposition to Antarctica over the past 34,000 years. *Nature*, Vol. 362, pp. 621-623 (1993).

Vosters, M., Contribution à la chimie des neiges Antarctiques. *Composition et origine des aérosols atmosphériques*. Thesis, Université Libre de Bruxelles, Belgium (1971).

Vosters, M., Hanappe, F. and Buat-Menard, P., Determination of Cl, Na, Mg, K and Ca in firn sample 66-A-6 from New Byrd Station, Antarctica-Comparison with work of Murozumi, Chow and Patterson. *Geochimica et Cosmochimica Acta*, Vol. 34, pp. 399-401 (1970).

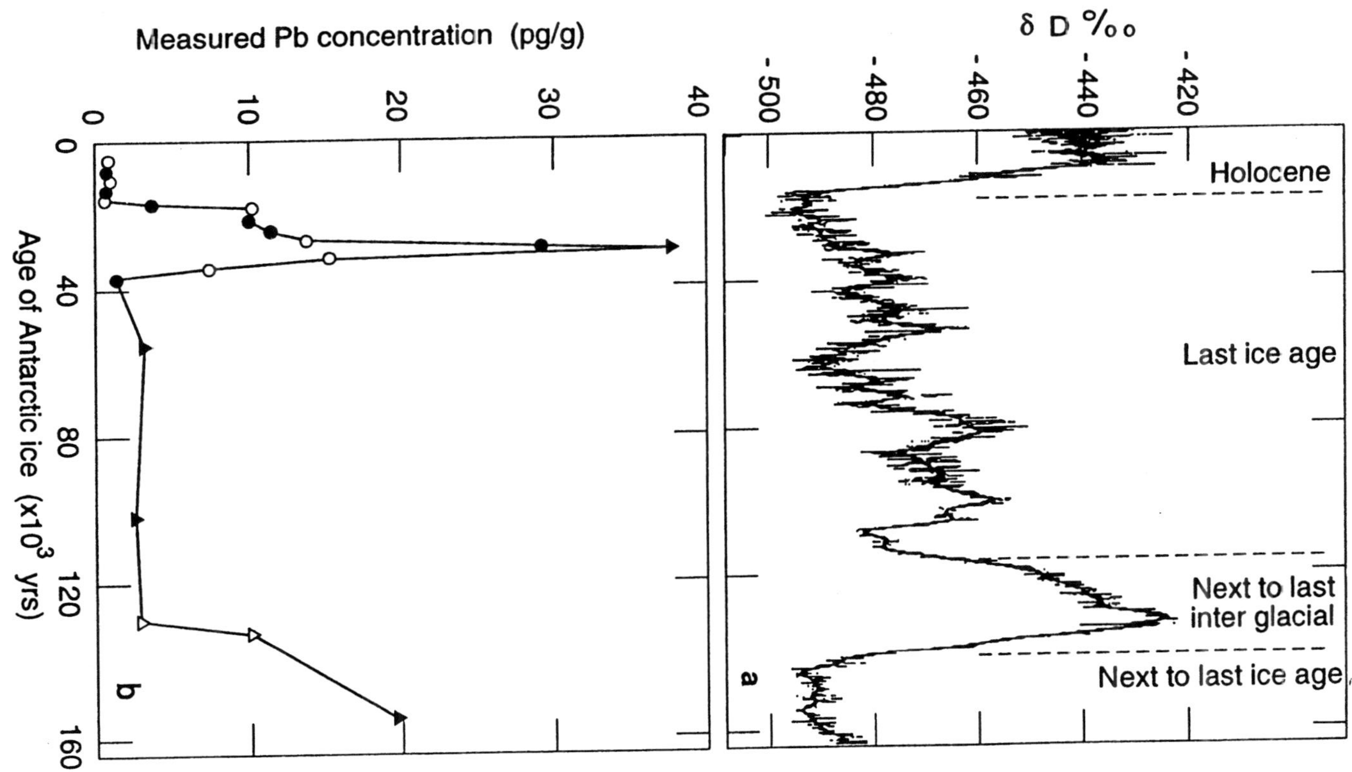

Figure 3. Changes in measured concentration as a function of radius in two sections of the 2083 m Vostok deep Antarctic ice core. The 26200 yrs BP (498.90 - 499.20 m deep) section was 108 mm in diameter. The 134000 yrs BP (1850.35 - 1850.54 m deep) section was 91 mm in diameter. From Boutron et al. (1987).

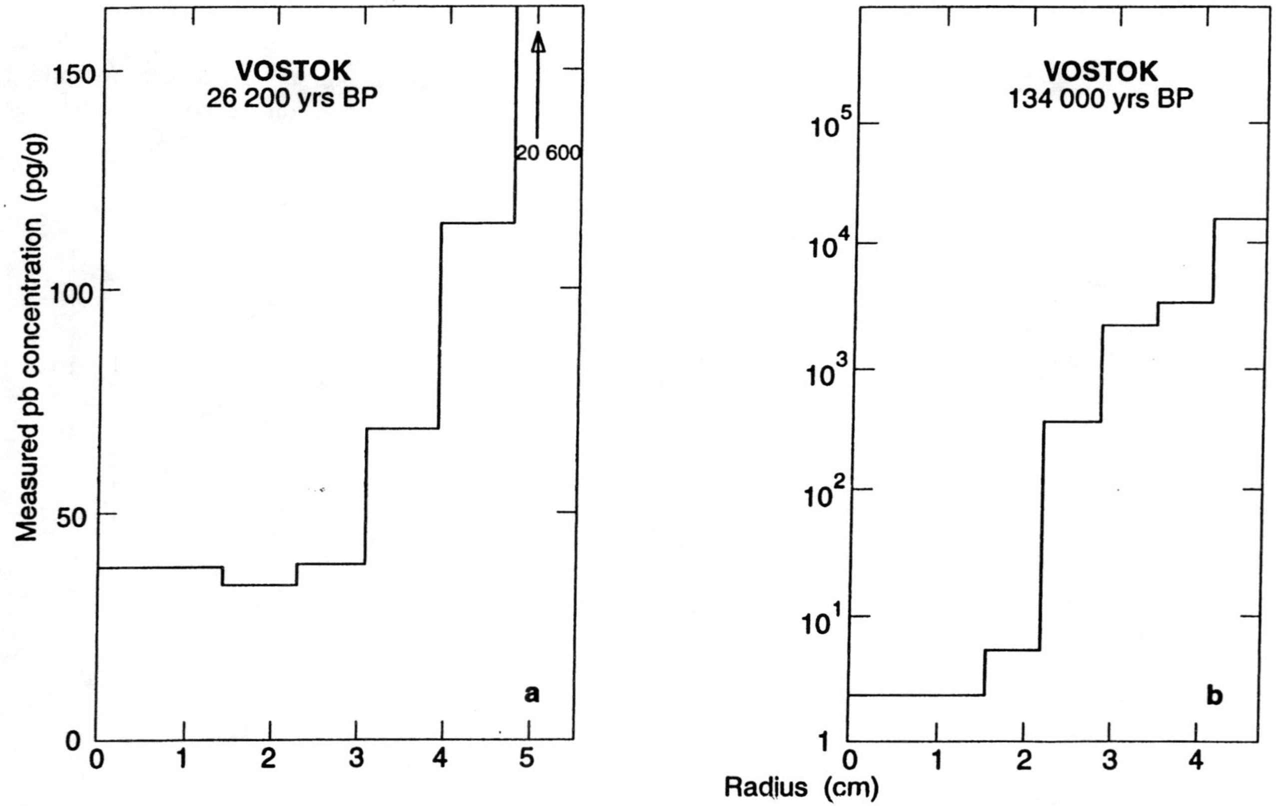

Figure 4. (a) Changes in climate during the last climatic cycle as documented through variations of δD‰ in the 2083 m Vostok Antarctic ice core (b) Changes in Pb concentrations in Antarctic ice during the last climatic cycle. Circles: Dome C data (Boutron and Patterson, 1986); triangles: Vostok data (Boutron et al., 1987). Black (full) symbols are used when a clear plateau of concentration was obtained on the outside-inside profile for the corresponding core section. Open symbols are used either when there was no plateau or when no outside-inside profile was determined.

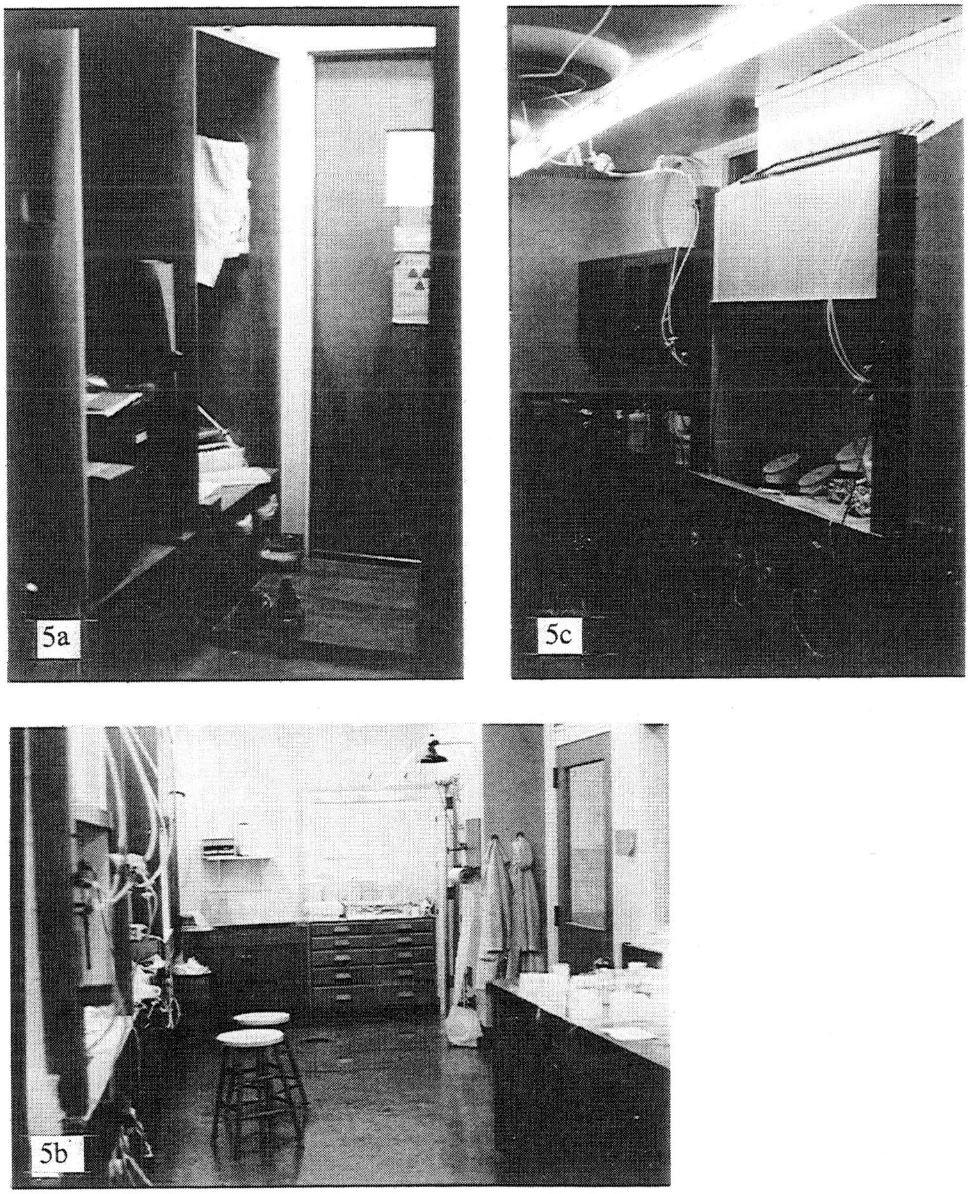

Figure 5. Clair Patterson's clean laboratory. (a) entrance of the laboratory. (b) general view. (c) one of the fume hoods. Photos by C. Boutron.

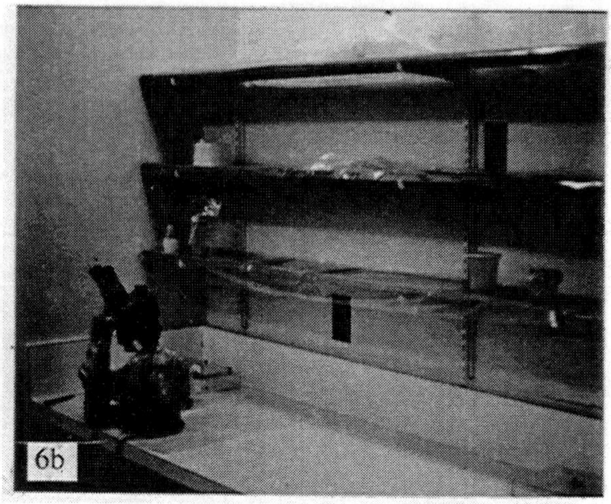

Figure 6. Clair Patterson's clean laboratory. (a) The still room. (b) The loading room. (c) The spikes. (d) Loading an Antarctic snow sample. Photos by C. Boutron.

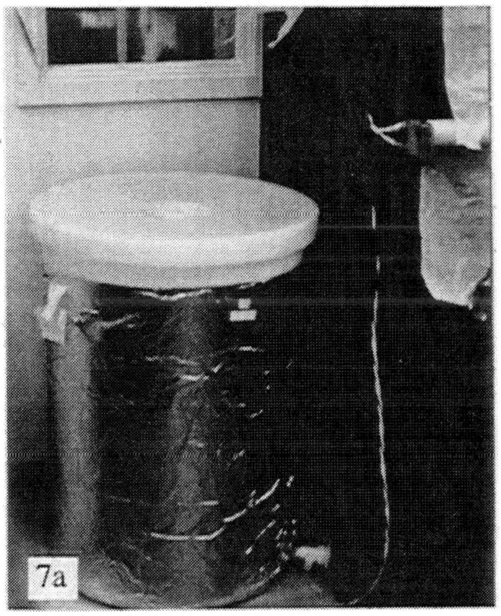

Figure 7. Clair Patterson's clean laboratory (a) One of the large acid baths. (b) The special medium size acid baths for the Antarctic snow and ice work. Photos by C. Boutron.

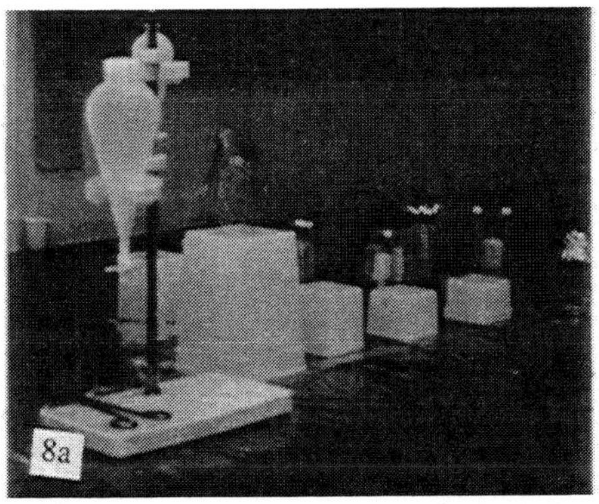

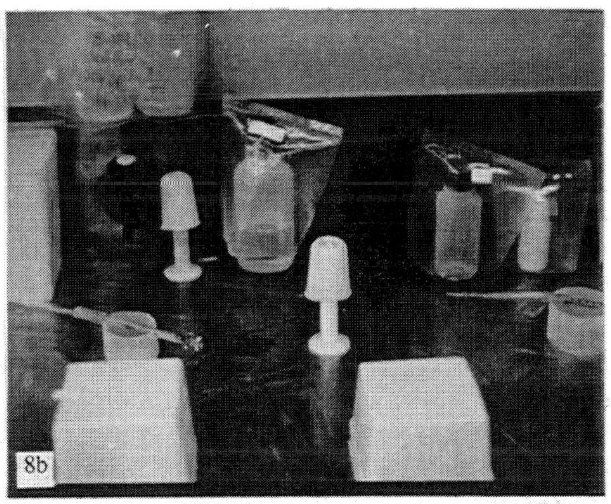

Figure 8. Clair Patterson's clean laboratory. (a) The extraction bench. (b) Close view of the quartz pipettes and the FEP teflon graduated vial. (c) One of the PTFE teflon nitrogen flushed chambers for the final evaporation of the extracted samples. Photos by C. Boutron.

CHAPTER 10

CLEAN HANDS, DIRTY HANDS: CLAIR PATTERSON AND THE AQUATIC BIOGEOCHEMISTRY OF MERCURY

WILLIAM F. FITZGERALD
Department of Marine Sciences
University of Connecticut
Groton, Connecticut 06340

INTRODUCTION

Clair Cameron "Pat" Patterson was my scientific father for more than twenty years. Our relationship began during a scientific meeting that was held at the Brookhaven National Laboratory in late May 1972. We were part of a working group of environmental and oceanographic scientists who had gathered to present and examine recent data for the amounts of heavy metals, halogenated hydrocarbons, and petroleum in seawater and biota from a variety of oceanic regimes. This workshop drew me to Pat.

Two events occurred between Thursday and Sunday to alter the course of my personal and professional life. The first I have called "The Patterson Challenge," which took place at Brookhaven, while the second was my marriage on May 28, 1972 to Patricia "Pat" Kelly Smith in Lafayette, Louisiana. Each wedding anniversary is a reminder of the continuing love and support that I have received from the Pats in my life. This memoir provides the opportunity to share my reflections on "The Patterson Challenge," and its influence in enhancing our understanding of the behavior and fate of mercury in the environment.

Brookhaven afforded Pat a bully pulpit for a "fire and brimstone" sermon directed at marine trace metal chemists. He struck fearlessly and profanely at presumptuous environmental scientists. Rhetorically, he banished the careless from the court of trace metal specialists to suffer in the purgatory for compromised work. Loosely paraphrased, Pat self assuredly charged, "Not one of you can measure Pb (and by implication other heavy metals) accurately in seawater because your methods are inaccurate." He elaborated, and following acrimonious debate, an angry challenged group agreed to an intercalibration exercise to be directed by Pat and his Caltech group. Predictably, we failed the challenge. However, with Pat's help and willingness to educate, we learned to measure Pb and other trace metals accurately in sea water. Pat outlined our education in a brief report, "Summary of Lead in Seawater Workshop," that appeared in Science (Patterson, 1974). The details were made available in a paper authored by the participants and published in Marine Chemistry (Berand et al., 1976).

As the Brookhaven Meeting and the intercalibration program demonstrate, Pat had influenced modern environmental studies of trace metals in a pivotal manner. This is especially evident in the marine environment, where his efforts bolstered the scientific framework for NSF's (National Science Foundation) sponsorship of pollutant-related studies in the oceans. Personally, it lead to our participation first in the Pollutant Transfer Program (1972-76) which was supported by NSF's Office for the International Decade of Ocean Exploration (IDOE), and then to the Sea-Air Exchange Program (SEAREX, 1977-87). Today, marine trace metal geochemistry is a well developed reaction and process oriented field. Thanks to Pat, we have progressed far beyond the cataloging of concentration data.

Interestingly, but not surprisingly, Pat was probably unaware of the significance and breadth of his contributions. True to his enigmatic and iconoclastic nature that is well-known to his friends and colleagues, Pat usually played down any suggestion that he had a significant impact on the study of trace metals in the environment. Yet he was keenly aware of his stature and influence. Even as a Professor Emeritus, he stood like the foreboding Giant in "Jack and the Beanstalk," ever self-assured and at the pinnacle of environmental studies. Indeed, we hoped Pat would continue indefinitely to rage and rule on the quality of data, screaming at the compromised efforts and interpretations offered by datagatherers.

Fortunately, Pat could be a patient mentor, the Henry Fonda look-alike oracle of experimental technique and scientific reasoning. Often, in this mentoring role, he passed on information in a theatrical manner and in the form of mini-dramas. I thought there would be interest and value in learning how his play "Clean Hands, Dirty Hands" fared in the heartland of this country, and how it influenced mercury studies on an international scale.

AMERICAN SAMOA, 1981

The story opens in American Samoa during January 1981. Pat, Dorothy Settle, and other environmental scientists from a variety of universities and research institutions were in Samoa as part of the Phase II experiments in SEAREX. I was serving as Co-Chief Scientist with Bob Duce, the moving force behind SEAREX, and the head of the University of Rhode Island investigations. During a decade of research, SEAREX scientists conducted intensive and pioneering air-sea exchange studies in the North Pacific tradewinds at Enewetak, Marshall Islands (1979) and in the South Pacific tradewinds at American Samoa during 1981. The South Pacific westerlies were investigated at Ninety-Mile Beach, North Island, New Zealand in 1983, while North Pacific westerlies were studied by ship in 1986. The studies were concerned with the sources, atmospheric transport, and air-sea exchange fluxes of inorganic and organic substances to the open ocean. The species included trace elements such as Pb, Hg, Se, Al, Fe, V, and rare earths; alkali and alkaline earths; sulfur compounds and nutrients; halogens; radionuclides (e.g., ^{210}Pb and ^{210}Po); natural organic compounds such as fatty acids, alcohols, aliphatic hydrocarbons and wax esters; synthetic organic substances (e.g., PCBs, DDT); and soil dust. The linkage between continental pollution, long range atmospheric transport, and the remote oceanic distribution of Pb and Hg in air, precipitation, and sea water provided a strong union between Pat and me. This metallic and environmental bonding, while not spark-free, has provided a foundation for my studies of Hg.

The January assembly in Samoa included SEAREX veterans of the Enewetak intensive campaign as well as a few new recruits. The experimental facility had been laboriously constructed on a rugged volcanic promontory of Cape Matatula at the eastern tip of the main island of Tutuilla. The aluminum tower, building materials, supplies, and sampling equipment had to be carried, hauled, pushed, and lugged from the staging site at the National Oceanic and Atmospheric Administration (NOAA) Geophysical Monitoring for Climatic Change (GMCC) Observatory that was situated majestically on a lofty cliff overlooking our station and providing a magnificent view of the vast Pacific. We all remember the climb, the 180 ladder-like stairs, and the Sisyphean struggle to reach the grassy level of the GMCC station. The resident complement of Samoan toads was frequently disturbed and astonished by over-heated, panting scientists carrying odd things while speaking strangely.

Most days were bright, warm, humid and typically windy at this beautiful South Pacific trades' site. NOAA rented the land from Chief Uli of Tula. The village of Tula was nestled on a carbonate berm between the upward sweeping lush green vegetation containing Tula's banana tree plantation and the bright coral beach. The beach sloped gently into the tranquil turquoise waters of the reef-protected lagoon. The village was most Eden-like, especially when viewed from the atmospheric sampling tower at a height of 40 meters and about a kilometer distance. To working scientists, however, paradise was often a refreshing Western Samoan brewed Valima beer. One could buy cold bottles at the "mom and pop" stores along the road or on draught at the bar in the Rainmaker Hotel! Most of us stayed at the Rainmaker which is situated on Pago Pago harbor in a non-paradise setting about 15 km west of Cape Matatula.

Leaving such pleasantries, let us return to the basaltic niche containing the SEAREX sampling towers and support buildings. I recall readily an afternoon shortly before sampling was to commence. A solemn Pat was instructing two geochemical novitiates in the Program on the proper way to collect atmospheric and precipitation samples for Pb and other metals. His teaching technique is best described as lawful and distinctively colorful. What it lacked in gentleness and student sensitivity was compensated for by the quality of the information and the logical scientific foundation on which it was based. Oh, yes, the rhetoric, as usual, was direct, profane, and intimidating. One listened attentively even while angry or just eavesdropping.

I am not certain if Pat would have recognized his teaching formula. It was as if the Samoan atmosphere or memories of Maugham's "Rain" brought out his skills as a dramatist. I was spellbound as he explained the intricacies, manipulations, and yes, the drama associated with uncompromised ultra-clean sampling of air and rain for trace substances. The discourse was in the fashion of a play, which I have titled, "Clean Hands, Dirty Hands." In general, two persons are needed to properly collect and process air, precipitation or water samples. In this two person scientific field "play," one actor is called "clean hands" and the other " dirty hands." Pat's script outlined scrupulously the details of ultra-clean procedures. Briefly, "clean hands" is the star, the collector of artifact-free samples for trace metals present at vanishingly small concentrations in air, precipitation, and natural waters. "Clean hands" handles only materials which have been carefully cleaned and prepared for sampling. This person must be extraordinarily cautious to avoid the environmental scientists' nemesis, the contagion of contamination.

"Dirty hands" is the supporting player, the person who opens and closes the curtain on an unusually important environmental production; yet his or her duties are similar to the ordinary activities of a careful laboratory scientist. That is, "dirty hands" ensures the integrity of the sampling by dealing skillfully with the peripherals and ancillary materials, such as protective outer coverings, sampler supports, tools, and sample containers. At no time does "dirty hands" touch any part of the apparatus used directly to collect and process the sample. For example, rain funnels and sampling bottles used for the accurate determination of Pb and Hg in open ocean precipitation must be prepared in an ultra-clean trace metal laboratory. The funnel and bottles are exhaustively acid-cleaned. The final stages of cleaning involve the highest quality sub-boiled distilled acids and waters. Ultra-clean materials are usually tripled-bagged in polyethylene, with the inner bags cleaned in acid. The outer bag is dirty since it is exposed to unclean situations ranging from shipping containers, benches, and sampling platforms, to dusty indoor air. "Dirty hands," as part of a well-rehearsed scientific sampling duet, enables "clean hands" to set up, handle, collect and process rain for later analysis. I was most impressed by how effectively Pat used a skit to define the responsibilities and choreograph the duties associated with the ultra-clean contamination-free collection of environmental samples. I even suggested to Pat that he might wish to put the production in writing.

NORTH CENTRAL WISCONSIN, 1986

The scene shifts by a few years and several thousand miles to rural north central Wisconsin and to human health concerns and sport fishery worries about elevated concentrations of Hg in freshwater fish. The state health advisory issued in 1986 (WDNR/WDH, 1986) reported that game fish (i.e., walleye pike) from about 30% of more than 300 lakes sampled contained Hg in amounts exceeding state health advisory limits (0.5 $\mu g\ g^{-1}$ wet weight). Many of the lakes were far from industrial regions and other localized sources of Hg pollution. Yet the Hg data obtained by the Wisconsin

Department of Natural Resources (WDNR) and its contractees for air, lakewater, and precipitation showed high Hg levels indicative of severe contamination. This inconsistency troubled Carl Watras, a limnologist with WDNR, who is stationed at the Trout Lake Laboratory of the Center for Limnology, University of Wisconsin. Carl sought me out and in a true Patterson fashion, I told Carl bluntly that the air, water, and precipitation data were "crap." Also, I explained that I was much too busy with our North Pacific SEAREX experiments to be of immediate help. Persistence and perseverance prevailed over my claims of overwork, and I agreed to collect and analyze a limited number of water and air samples.

The experimental design was simple. The initial goal was to determine Hg in samples that would be representative of average lakewater and could be obtained under uncompromised trace-metal free conditions. We would provide a benchmark for Hg in clear-water lake types ranging from small, dilute seepage lakes whose hydrologic budget was dominated by precipitation to a large drainage lake. The study lakes were chosen to provide a gradient in pH and acid neutralizing capacity, two parameters that appeared linked to the bioaccumulation of Hg.

Our studies of Hg in the near shore environs of Long Island Sound and Narragansett Bay suggested that amounts of Hg in these rural Wisconsin lakes would be small. Accordingly, we collected and analyzed samples using trace metal-clean techniques that we developed, under Pat's tutelage, to study the biogeochemical cycling of Hg in remote, pristine oceanic environments (Gill and Fitzgerald, 1985, 1987). Thus, the lakes were sampled after fall overturn (autumn mixis) so that surface collections would provide a measure of the average Hg concentration for each lake. Surface water was carefully sampled by hand from a fiberglass workboat. The collections were made off the bow as the boat was moved gently forward by a battery-powered outboard motor. I wore arm-length polyethylene gloves to collect water from ca. 0.5 meters in 2L acid-cleaned Teflon bottles. Jon Kim, graduate student and a Samoan veteran, acted as 'dirty hands' and assisted with collection, field acidification, hermetic sealing, and bagging of the samples. These were later analyzed at our laboratories in the Department of Marine Sciences at the University of Connecticut. We are indebted to Grace Vandal and Nicolas Bloom for their careful work.

At the time of the Wisconsin lake survey, we routinely made two operationally defined measurements of Hg in natural waters, total Hg (Hg_T) and reactive Hg (Hg_R). Total Hg is determined following strong oxidation of an unfiltered water sample using bromine-monochloride, while Hg_R represents those Hg species that are readily reducible (e.g., $SnCl_2$) in aliquots acidified to a pH of ~ 1. The acid-labile fraction, Hg_R, will include dissolved inorganic species, some organo-Hg associations, Hg that is readily leached from particulate matter, and any elemental Hg present. The Hg_T measurement includes the Hg_R components as well as very stable organo-Hg associations such as methyl-Hg. With the exception of low-salinity estuarine environments, we had found little difference between Hg_T and Hg_R in marine studies.

RESULTS

Results from the preliminary baseline investigation were most surprising and of much significance in challenging the prevailing view of the factors controlling the behavior, fate, and bioaccumulation of Hg in lacustrine systems. The data, summarized in Figure 1, indicated unequivocally that prior studies were in gross error. While we anticipated Hg in the selected northern Wisconsin lakes to be lower than previously reported, we were astonished to find Hg_T and Hg_R levels approaching those of remote ocean water. Contrast the range of Hg_R (0.7 to 2.9 pM) and Hg_T (4.7 to 9.7 pM) observed in our survey with previous Hg data for these lakes which ranged from 90 to > 1000 pM (Glass et al., 1986)! The other investigators lacked the geochemical trace metal acumen and ultra-clean sampling and analysis skills needed to study Hg in the environment. They had not been tutored by Pat Patterson. Carl Watras and I illustrated the extraordinary poor quality of the past studies by use of a time series for one of the lakes. The historical pattern of Hg concentrations measured in Vandercook Lake is reproduced in Figure 2.

As illustrated, there was a dramatic decline in the Hg concentrations in Vandercook Lake when we applied stringent trace metal-clean techniques for the sampling and analysis of Hg in lake waters. The true Hg levels correspond more closely to our seawater measurements (1 to 11 pM) at that time (Gill and Fitzgerald, 1988). Moreover, the difference between our determinations (Hg_R= 2.5 ± 0.2 pM) and the two previous studies in 1983 (Hg_R = 1200 ± 600 pM) and 1985-86 (Hg_R = 300 ± 150 pM) was almost beyond belief. However, we had seen a similar historical pattern with a variety of trace metals in the marine environment as improvements in sample collection and handling techniques were introduced. Figure 3 was used to illustrate the substantial decreases in mean oceanic concentrations of several representative trace metals that had been measured over thirty years.

Not surprisingly, Pat loved both of these plots. The inaccurate picture provided by careless investigators was exposed. This evidence shocked and convinced recalcitrant environmental scientists and managers that the Hg problem in lakes was subtle and complex.

The results are linked to the Patterson Challenge, and the continuing harvest of benefits in diverse environmental investigations that are derived from Pat's studies of Pb and his willingness to share expertise and to educate. Here, the revolution in oceanic trace element research has spread to terrestrial studies of Hg. Further, and as plotted in Figure 1, a measure of the complexity is evident. The difference between Hg_T and Hg_R indicated that a significant amount of the Hg in lake water was organically associated. Subsequent work has shown that the organo-Hg associations are not due principally to methylated Hg species, but most probably involve Hg-humic associations. Today, there is much interest in deciphering the complex role such organo-Hg species play in affecting the bioavailability and bioaccumulation of Hg in freshwaters.

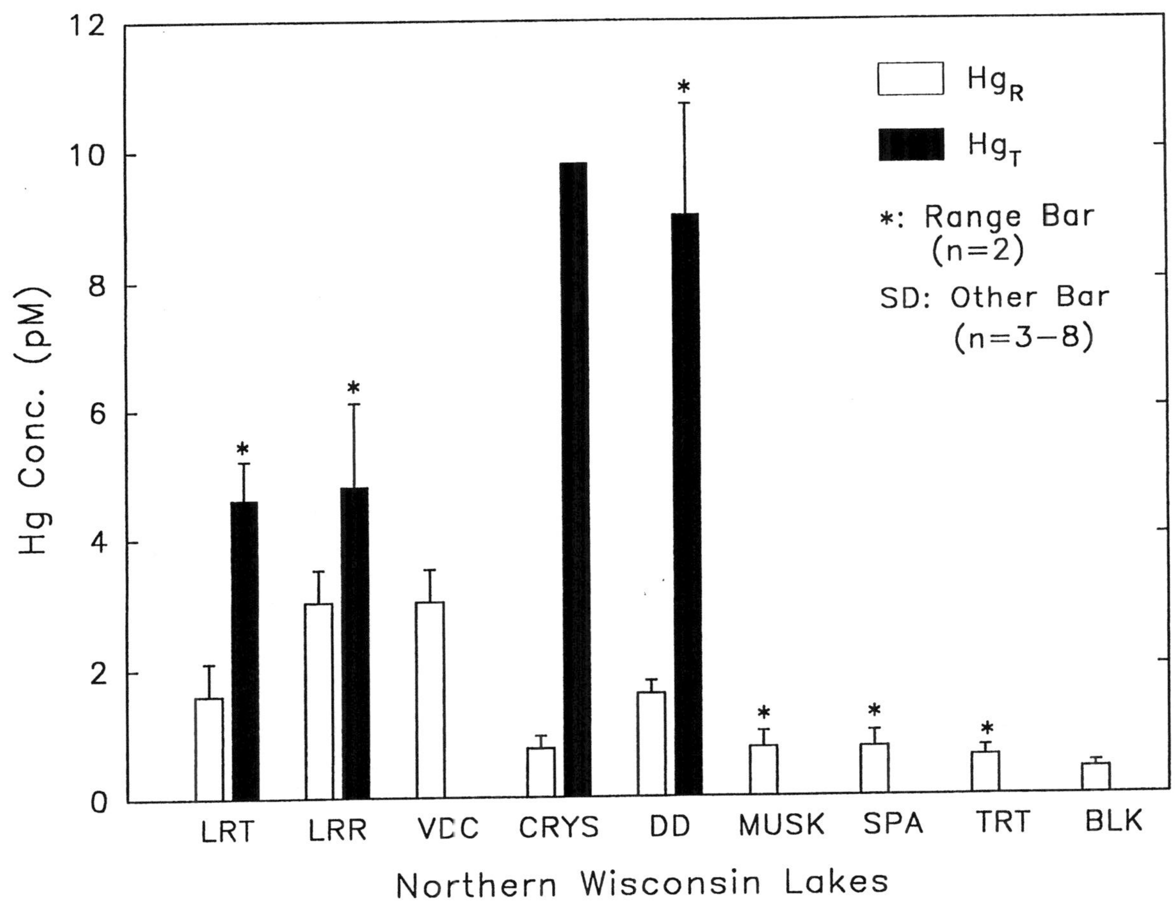

Figure 1. Picomolar (pM) concentrations of total and reactive Hg in unfiltered surface waters from lakes in rural northern Wisconsin. The lake abbreviations refer to: Little Rock, treatment and reference basins; Vandercook; Crystal; Dorothy Dunn; Big Muskellunge; Sparkling and Trout. The blank is indicated as BLK and it was about the same for both the total and reactive Hg determinations. This illustration has been redrawn by Carl Lamborg from Fitzgerald and Watras (1989) with permission.

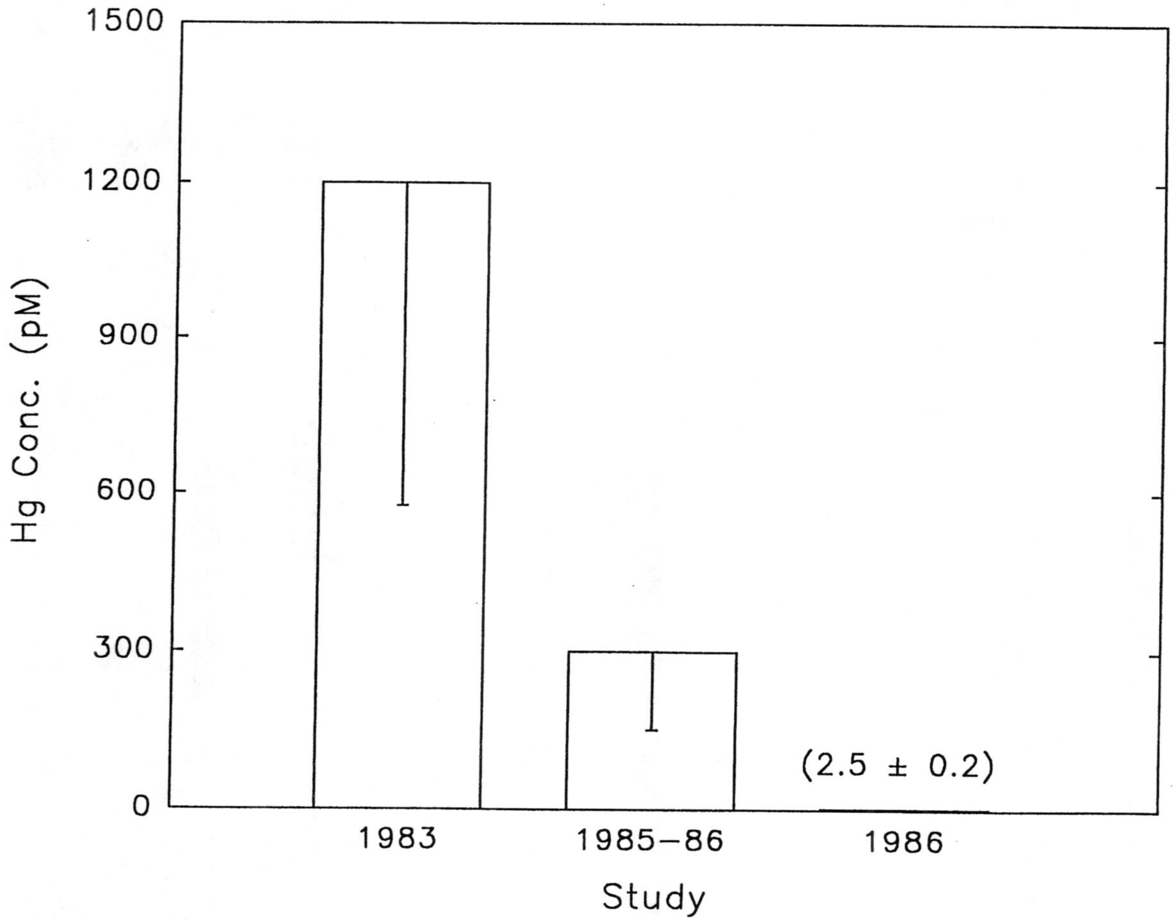

Figure 2. Three different studies of mercury in surface waters of Vandercook Lake, Wisconsin. The measured concentrations decreased with the adoption of progressively "cleaner" sample collection and analysis techniques, that culminated with the Patterson approach employed by the author in 1986. This illustration has been redrawn by Carl Lamborg from Fitzgerald and Watras (1989) with permission.

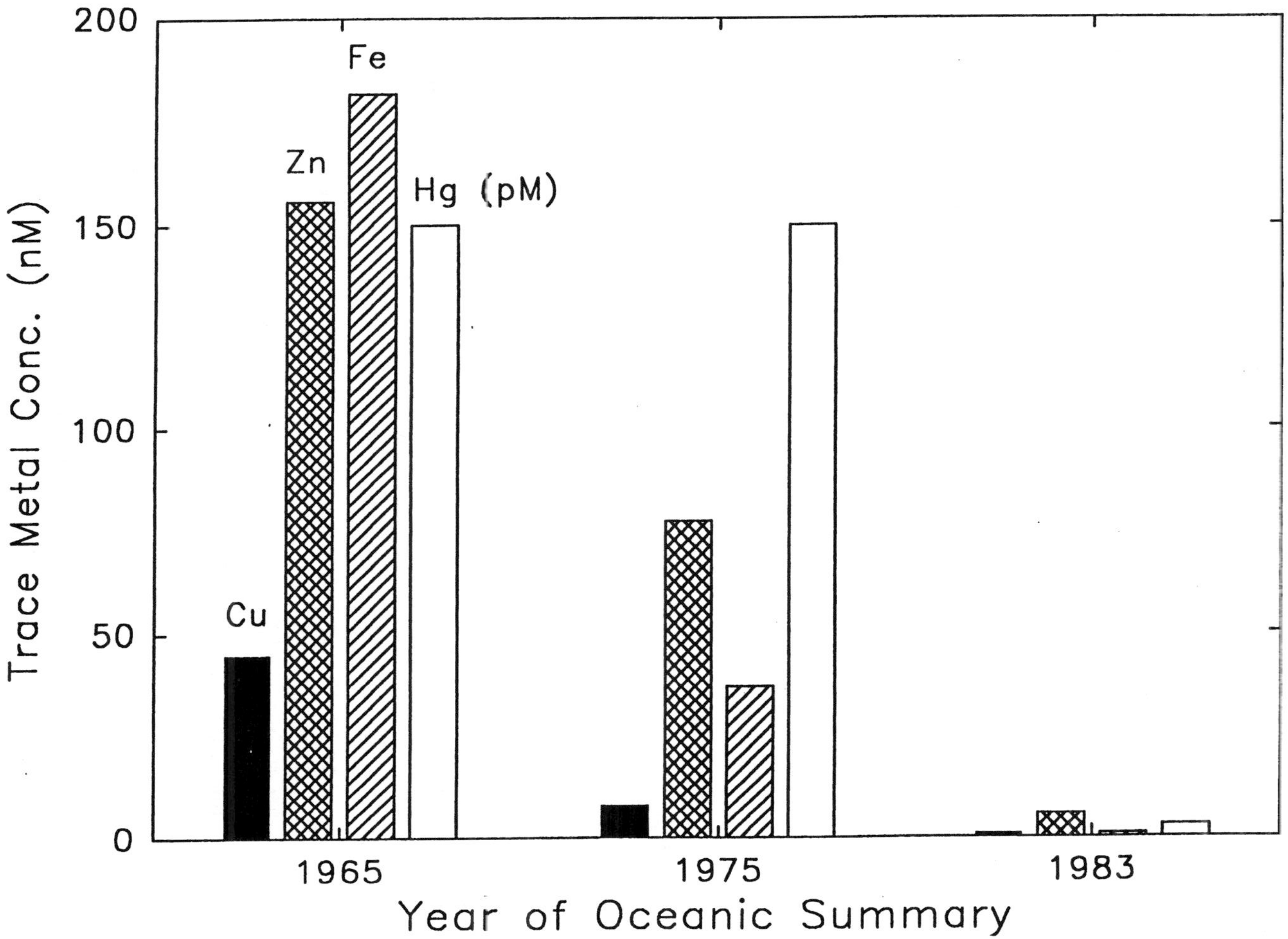

Figure 3. Mean oceanic concentrations for several metals (Cu, Zn, Fe, and Hg) as estimated in reviews by Goldberg (1965), Brewer (1975) and Bruland (1983). This illustration has been redrawn by Carl Lamborg from Fitzgerald and Watras (1989) with permission.

We made a few measurements of atmospheric Hg and found an average of 2.0 ± 0.2 ng m^{-3} for the total gaseous Hg (TGM). This concentration is similar to the values we observed over the North Central Pacific Ocean where the TGM varied between 1.5 and 2.2 ng m^{-3} with a mean and standard deviation of 1.77 ± 0.15 ng m^{-3} (Fitzgerald, 1989). We were not surprised by this correspondence between the mid-continent and the open ocean, since TGM consists almost entirely of elemental Hg (>95%). The similarity is indicative of the relatively long atmospheric residence time of Hg^{o} (~ 1 year) which allows natural and anthropogenic emissions of Hg^{o} to mix readily throughout the hemisphere, thereby affecting open ocean regions. Further, the agreement of TGM between marine and terrestrial systems suggested that the concentrations of Hg in the air in particulate form (Hg_P) and in precipitation from the Wisconsin study region would not be greater than our prior determinations of these constituents in the environs of Long Island Sound. We had found Hg_P to range between 8 and 60 pg m^{-3} (Fitzgerald, et. al., 1983). In rain, the mean and standard deviation for Hg_R was 50 ± 25 pM (Fogg and Fitzgerald, 1979). As with the lake waters, these data were considerably lower than the values being used by the WDNR in 1986.

FIRST ORDER MODELING

As Pat would emphasize, there are two tests of data quality. First, the accuracy of the analytical determinations must be documented and frequently demonstrated using laboratory standards and reference materials. The quality control and assurance checks must include the sampling procedures. Scrupulous attention to the preparation of ultra-clean collection apparatus and the maintenance of its integrity during use represents the essence of excellent field work. The second test of quality is the requirement that the data fit a coherent environmental picture. Accordingly, and although we were working with a limited amount of data, we tested the Hg data for consistency by producing a preliminary coarse mass balance for Hg in the one of the lakes, Little Rock Lake (LRL).

Little Rock Lake is a clear water seepage lake that had been divided with a sea curtain into two basins, one which was untreated and maintained as a reference (LRLR), and the other which was bring experimentally acidified, referred to as the treatment basin (LRLT). These basins had been extensively studied and reliable data existed for Hg in the fish stock (Cope et al., 1990) and in the sediments (Rada et al., 1989), while rates of sedimentation were available from Tacconi (1988). The preliminary Hg budget for LRLT is presented in Figure 4. Details of this work can be found in Fitzgerald and Watras (1989) and in Wiener et al. (1990).

Briefly, the Hg inputs could be restricted to wet and dry deposition since LRL receives > 95% of its hydrological inputs as direct precipitation. Mercury losses were limited to net sedimentation which was estimated as the product of annual deposition and the Hg mass in the sediment material that was available from investigations by Ron Rada

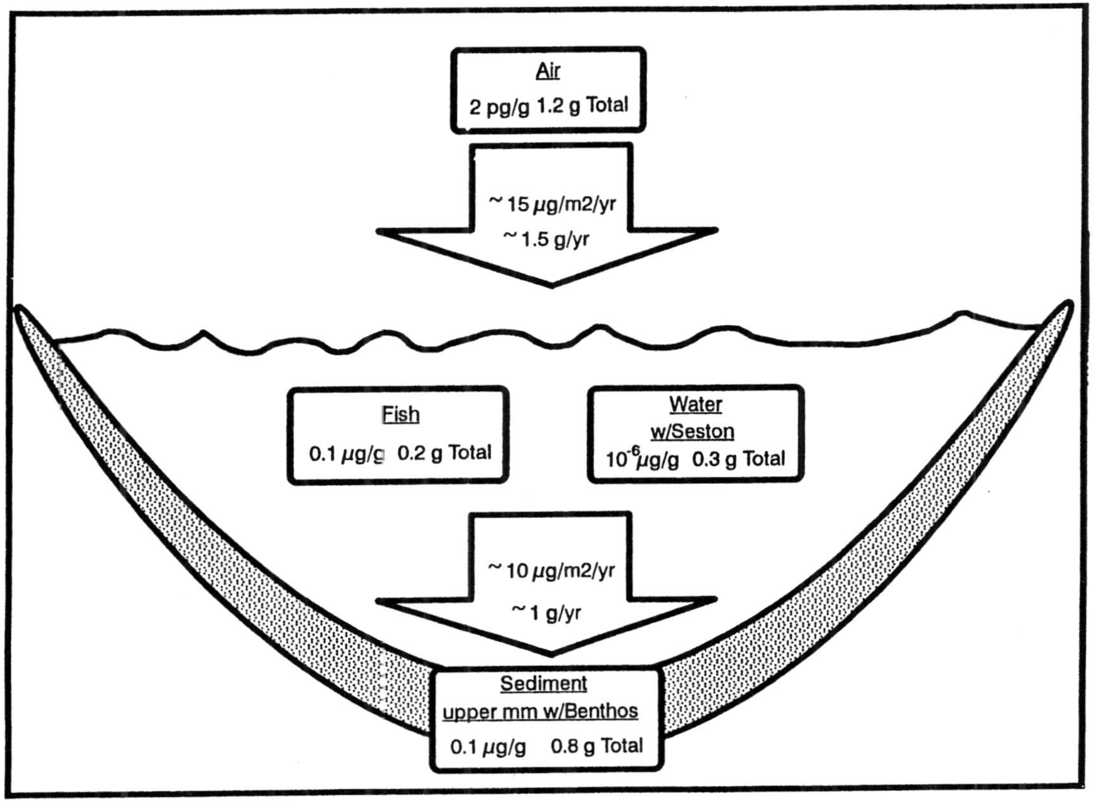

Figure 4. Preliminary mass balance for Hg for the treatment basin of Little Rock Lake, Wisconsin. The illustration has been adapted from Fitzgerald and Watras (1989) and Wiener et al. (1990). Redrawn by Carl Lamborg from Fitzgerald and Watras (1989) with permission.

of the University of Wisconsin at La Crosse. The amounts of Hg in fish were derived from research conducted by Jim Wiener's group from the US Fish and Wildlife Service, La Crosse. The atmospheric deposition flux of Hg was estimated to be 15 $\mu g\ m^{-2}\ year^{-1}$ by assuming that it would be about 50% greater than our estimates of Hg deposition (11 $\mu g\ m^{-2}\ year^{-1}$) in the environs of Long Island Sound (Fitzgerald, 1986). Although many Hg species and the complexity of the aquatic biogeochemical cycling of Hg were ignored, this mass balance yielded an accurate and consistent simple model for the distribution and fluxes of Hg in a temperate seepage lake. Moreover, and because the quality control constraints of accuracy and consistency were met, we used the mass balance to place various processes affecting the behavior and fate of Hg in these lakes in the proper environmental perspective. This work provided a scientific framework for follow-up research in this region as well as a reference for other studies.

The budget showed that relatively small amounts of atmospheric Hg deposition would readily account for the total mass of Hg in fish, water, and accumulation in the sediments of Little Rock Lake. Indeed, just 1.5 grams of Hg, the amount in a fever thermometer, appeared to be entering the basin each year. This suggested that increases in the atmospheric Hg loading could lead directly to increased Hg levels in fish. We noted, for example, that by assuming the turnover of fish biomass was 30% $year^{-1}$, the amount of monomethylmercury (MMHg) needed to account for the annual accumulation would be just 0.05 g, or only about 3% of the estimated atmospheric Hg deposition. It would be necessary to convert only a small amount of the atmospheric Hg input to MMHg to provide the MMHg found in fish. The airborne Hg contamination in this lake region was subtle rather than gross, and the aquatic biogeochemical cycle of Hg was quite different than the erroneous views that prevailed before this study.

The mass balance showed that the insidious nature of the Hg cycle leads to the presence of elevated levels of MMHg in fish. It was evident that the environmental behavior and accumulation of mercury in aquatic organisms was subtly complex and driven by chemical and biologically mediated reactions involving exceedingly small quantities of mercury in the atmosphere and natural waters. This environmental revelation pointed to the importance of an ultra-trace atmospheric and aquatic biogeochemical mercury cycle which includes non-point sources and long range airborne transport that had been overlooked in earlier research. While the broad atmospheric dispersion of Hg^{o} probably reduced some localized impact from human related emissions, it may have led to the geographically large problem of elevated methylmercury concentrations in fresh water and marine fish that are far removed from local sources.

The severe Hg problems of the late 1960's and early 1970's were thought to be due completely to acute point sources of mercury contamination. The problems were assumed to be solved following the identification and the elimination or reduction of discharges from large sources of mercury pollution. Knowledgeable researchers, however, were quite aware that the environmental behavior and fate of mercury was poorly understood; many inconsistencies were apparent, and the root causes of elevated mercury concentrations in aquatic organisms had not been established. For example, no one offered a satisfactory explanation for the elevated quantities of mercury in fish from oceanic regions far from local sources of mercury. Similarly, the situation in these semi-remote mid-continental lakes appeared to be related to non-point source emissions and atmospheric Hg transport and deposition. It must be emphasized that a few accurate measurements of Hg in Wisconsin lakes and some reasonable geochemical scaling using reliable Hg data from marine regions allowed a simple mass balance to be generated. This first -order but accurate budget put the situation in a reasonably well-constrained perspective.

MODEL REFINEMENTS

This new perspective on the Hg cycle in lakes led to broad multidisciplinary investigations of the aquatic biogeochemistry of Hg. The initial program was entitled Mercury in Temperate Lakes (MTL), and it was quite successful. The research continued until 1996 as the Mercury Accumulation, Pathways and Processes (MAPP) program. Both programs were supported jointly by the Electric Power Research Institute and the Wisconsin Department of Natural Resources. Not unexpectedly, the results from our preliminary survey have been completely confirmed. A more refined budget for LRLT is presented in Figure 5. Briefly, as our initial mass balance suggested for this model temperate seepage system, LRLT, the total net atmospheric Hg deposition (Hg_T) of ca. 10 $\mu g\ m^{-2}\ y^{-1}$ (ca. 66% wet) readily accounts for the total mass of Hg in fish, water, and accumulation in the sediments. The ecosystem appears delicately poised with respect to atmospheric inputs, since a relatively small fraction of the input readily accounts for the estimated accumulation of Hg in fish. The average Hg concentration for the lake is identical to the estimate we made based on the two surface water values that were collected following the autumn mixis. The fish and sediment data are approximately the same. There are no substantial changes in the coarse view of the fluxes and amounts of Hg in various compartments.

The linkage between increases in atmospheric Hg loading and elevated levels in the fish stock is evident. Moreover, this connection has been confirmed in a mechanistic model of the biogeochemical cycling of mercury in lakes that was developed by Hudson et al. (1995) using data provided by investigators in the MTL study. There were many additional improvements in knowledge of the biogeochemical cycling of Hg in natural waters that have been derived from these seepage lake studies, which began with a Patterson-style experiment. The work has progressed from a coarse framework to detailed biogeochemical investigations and modeling. Mercury species, especially MMHg, are routinely measured, thereby allowing a direct examination of individual components of the Hg cycle. For example, atmospheric deposition of MMHg is insufficient to account for the amounts of MMHg observed in biota, as shown in

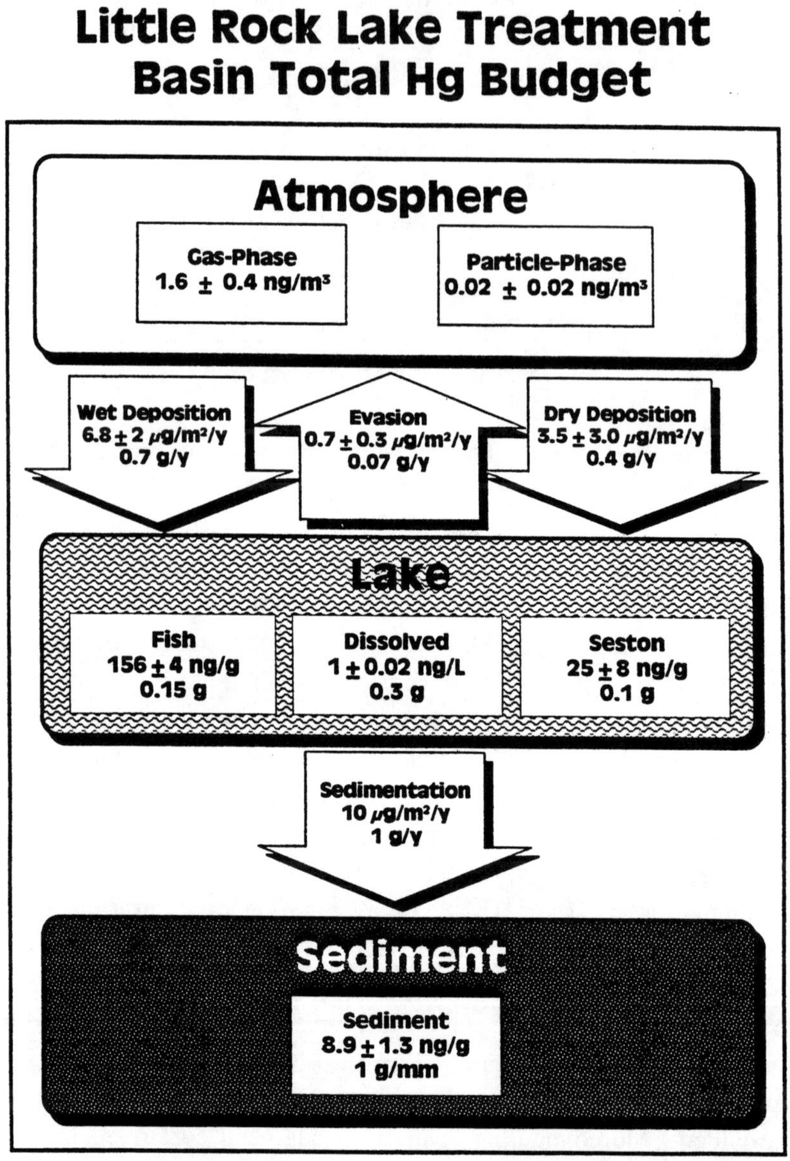

Figure 5. Mass balance model for Hg in the treatment basin of Little Rock Lake, Wisconsin (adapted from Fitzgerald et al., 1991 and based on work and preliminary budgets appearing in Wiener et al., 1990, Fitzgerald and Watras, 1989, and Watras et al., 1994). Redrawn by Carl Lamborg from Watras et al. (1994) with permission.

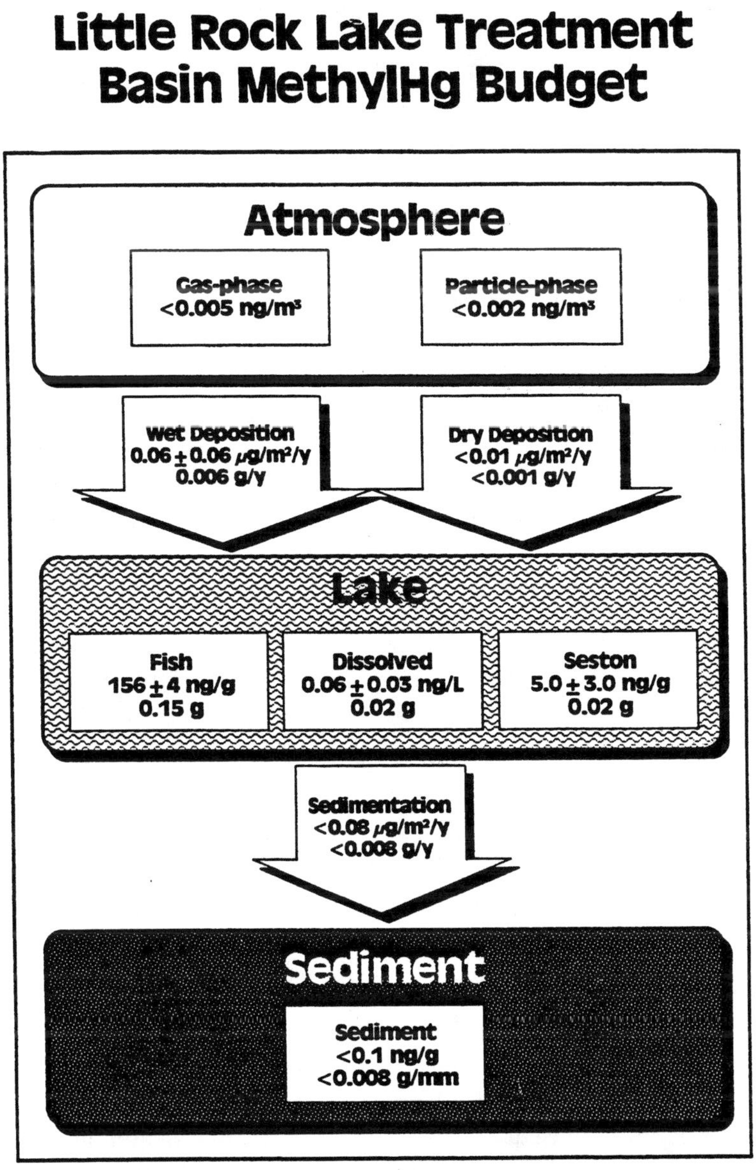

Figure 6. Mass balance model for monomethyl Hg in the treatment basin of Little Rock Lake, Wisconsin (adapted and based on work from Fitzgerald et al., 1991; Wiener et al., 1990; Hurley et al., 1991; Bloom et al., 1991; and Watras et al., 1994). Redrawn by Carl Lamborg from Watras et al. (1994) with permission.

Figure 6. This confirmed the need for an in-lake synthesis of MMHg. In-situ methylation is now well-documented for these temperate lakes as well as for a variety of other fresh water systems.

Using our marine work as an analog, we discovered an important in-situ production and evasion cycle of Hg^{o} in these seepage lakes (Vandal et al., 1991). Interestingly, once Hg^{o} is produced in the aqueous phase, it appears to be unreactive and will be eventually lost to the air. We have suggested that lakes with limnological conditions favoring Hg^{o} production are less likely to have elevated levels of methylmercury in the fish stock, and show smaller accumulation of mercury in the sediments. This hypothesis is being carefully examined (e.g., Rada et al., 1993).

The MTL and MAPP Programs have improved our understanding of the basic science, environmental effects, human health effects and management of Hg. These programs have had a major influence on current world-wide studies of Hg. The various investigative groups in MTL and MAPP have included a mix of scientists from state and federal agencies, private firms, and universities. In addition, other scientists from the US, Canada, and Scandinavia were brought aboard as advisors and collaborators. Intercalibration programs were initiated and Patterson's message of uncompromised high quality environmental work was spread. The “Clean Hands, Dirty Hands” system was embraced, mimicked, embellished, and even co-opted as the proper and only way to study Hg in the environment. It was as if the oceanic equivalent of sea-floor spreading had been discovered in terrestrial aquatic systems.

Our proselytizing efforts were enhanced on a global scale by the large number of contributions and fervor of the MTL investigators at the first international meeting dealing with Hg in the environment. This was held in Gävle, Sweden (1990), and the proceedings published in a special volume of Water, Air, Soil Pollution (W.A. S. P., 1992). Two years later, the second annual meeting was convened in Monterey, California (1992). There were twice as many participants as there were at the Swedish assembly. Subsequent conferences in Vancouver, British Columbia (1994), and in Hamburg, Federal Republic of Germany (1996) each attracted more than 500 attendees.

The interest in Hg is a consequence of the current human health and environmental concerns associated with elevated levels of methylmercury in freshwater and marine piscivorous fish. The rewards of the “Patterson Challenge” are evident on a world-wide basis. The confused picture of Hg cycling of a few years ago has been replaced by a high-quality database that includes extraordinary determinations of Hg species such as MMHg, DMHg, and Hg^{o}. Indeed, it appears that modern environmental and ecological studies of Hg are leading the way for other investigations of trace toxic constituents. For example, new analytical and geochemical information is being communicated effectively, especially by the biannual international meetings. Environmental Hg research is improving steadily and knowledge of the biogeochemical cycling of Hg is growing significantly. There is a new wave of exciting and important environmental mercury studies. Critical research areas include: (1) establishing patterns of modern and historic Hg deposition to provide an essential foundation for detailed biogeochemical and ecological studies of Hg, (2) assessing the contributions of global versus local and regional Hg sources to terrestrial and oceanic regions, (3) identifying the reactions associated with cycling of elemental mercury (Hg^0) in the atmosphere and natural waters, (4) investigating the mechanisms leading to the post-deposition in-situ bacterial conversion of mercury species to methylated forms, and (5) relating human exposure to methylmercury with the levels of lowest effect.

In summary, as we have seen, erroneous environmental data can have far reaching adverse effects. Pat's long battle to place the scourge of Pb pollution in an accurate framework in the face of huge quantities of inaccurate but published Pb data testifies to the confusion generated by uncaring and thoughtless investigators. Pat's anger and lack of patience with the perpetrators appear quite rational when one views the myriad of problems inaccurate data creates for scientists, regulators, planners, and human health advocates. Confusion and costly misdirection of effort have been a feature of many environmental efforts, especially with Pb and Hg. Improvements in our current understanding of Hg can be traced in part to Pat and his efforts to study the environmental and the anthropogenic impact of Pb at the highest level and in an open, uncompromised way. Merci, mon ami et mentor!

REFERENCES

Berand, P., Bewers, J.M., Duce, R., Fitzgerald, W.F., Hartling, S., Hunt, C. D., Patterson, C., Settle, D., Smith, R., Walsh, P., Windom, H. and Wong, C.S., Comparison determinations of lead by investigators in both their home laboratory and in an isotope dilution laboratory. *Marine Chemistry*, Vol. 4, pp. 389-392 (1976).

Bloom, N.S., Watras, C.J. and Hurley, J. P., Impact of acidification on the methylmercury cycle of remote seepage lakes. *Water, Air and Soil Pollution*, Vol. 56, pp. 477-491 (1991).

Brewer, P.G., Minor elements in sea water. In: *Chemical Oceanography*, J.P. Riley and G. Skirrow (Eds.), Academic Press, New York, pp. 415-496 (1975).

Bruland, K.W., 1983 Trace elements in sea water. In: *Chemical Oceanography*, J.P. Riley and G. Skirrow (Eds.), Academic Press, New York, pp. 157-215 (1983).

Cope, W.G., Wiener, J.G. and Rada, R.D., Mercury accumulation in yellow perch in Wisconsin seepage lakes. *Environmental Toxicology and Chemistry*, Vol. 9, pp. 931-940 (1990).

Fitzgerald, W.F., Gill, G.A. and Hewitt, A., Air-Sea exchange of mercury. In: Trace Metals in Seawater, C.S. Wong, E. Boyle, K.W. Bruland, J.D. Burton, and E.D. Goldberg, (Eds.), NATO Conference Series, IV, *Marine Science*, Vol. 9, Plenum Press, New York, pp. 297-316 (1983).

Fitzgerald, W. F., Cycling of mercury between the atmosphere and oceans. In: *The Role of Air-Sea Exchange in Geochemical Cycling*, P. Buat-Ménard (Ed.), D. Reidel Publishing Company, pp. 363-408 (1986).

Fitzgerald, W.F., Atmospheric and oceanic cycling of mercury, In: *SEAREX Volume of the Chemical Oceanography Series*, J.P. Riley and R. Chester (Eds.), Academic Press, London, pp. 151-186 (1989).

Fitzgerald, W.F. and Watras, C.J., Mercury in surficial waters of rural Wisconsin lakes. *Science of the Total Environment*, Vol. 87/88, pp. 223-232 (1989).

Fitzgerald, W.F., Vandal, G.M., and Mason, R.P., Atmospheric cycling and air-water exchange of mercury over mid-continent lacustrine regions. *Water, Air and Soil Pollution*, Vol. 56, pp. 745-767 (1991).

Fogg, T.R. and Fitzgerald, W.F., Mercury in southern New England coastal rains. *Journal of Geophysical Research*, Vol. 84, pp. 6987-6988 (1979).

Gill, G.A. and Fitzgerald, W.F., Mercury sampling of open ocean waters at the picomolar level. *Deep Sea Research*, Vol. 32, pp. 287-297 (1985).

Gill, G.A. and Fitzgerald,W.F., Picomolar mercury measurements in seawater and other materials using stannous chloride reduction and two-stage gold amalgamation with gas phase detection. *Marine Chemistry*, Vol. 20, pp. 227-243 (1987).

Gill, G.A. and Fitzgerald, W.F., Vertical mercury distributions in the oceans. *Geochimica et Cosmochimica Acta*, Vol. 52, pp. 1719-1728 (1988).

Glass, G.E., Leonard, E.N., Chan, W.H. and Orr, D.B., Airborne mercury in precipitation in the Lake Superior region. *Journal of Great Lakes Research*, Vol. 12, pp. 37-51 (1986).

Goldberg, E.D., Minor elements in sea water. In: *Chemical Oceanography*, J.P. Riley and G. Skirrow (Eds.), Academic Press, New York, pp. 163-196 (1965).

Hudson, R.J. M., Gherini, A.S., Fitzgerald, W.F., and Porcella, D.B., Perturbation of the global mercury cycle: A model-based analysis. *Water, Air and Soil Pollution*, Vol. 80, pp. 265-272 (1995).

Hurley, J.P., Watras, C.J. and Bloom, N.S., Mercury cycling in northern Wisconsin seepage lakes: the role of particulate matter in vertical transport. *Water, Air and Soil Pollution*, Vol. 56, pp. 543-551 (1991).

Patterson, C.C., Lead in seawater, Meeting Report. *Science*, Vol. 183, pp. 553-554 (1974).

Rada, R.B., Wiener, J.G., Winfrey, M.R., and Powell, D.E., Recent increases in atmospheric deposition of mercury to northcentral Wisconsin lakes inferred from sediment analysis. *Archives of Environmental Contaminant Toxicology*, Vol. 18, pp. 175-181 (1989).

Rada, R.G., Powell, D.E., and Wiener, J.G., Whole-lake burdens and spatial distribution of mercury in surficial sediments in Wisconsin seepage lakes. *Canadian Journal of Fisheries and Aquatic Science*, Vol. 50, pp. 865-873 (1993).

Tacconi, J. E., Ion budgets and cycling processes in an acid sensitive seepage lake. MS Thesis, University of Minnesota, Minneapolis (1988).

Vandal, G.M., Mason, R.P. and Fitzgerald, W.F., Cycling of volatile mercury in temperate lakes. *Water, Air and Soil Pollution*, Vol. 56, pp. 791-803 (1991).

Water, Air, and Soil Pollution, Vol. 56 (Special Issue), Mercury as an environmental pollutant, p. 847 (1992).

Watras, C.J., Bloom, N.S., Hudson, R.J.M., Gherini, S.A., Munson, R., Claas, S.A., Morrison, K. A., Hurley, J., Wiener, J.G., Fitzgerald, W.F., Mason, R., Vandal, G., Powell, D., Rada, R., Rislow, L., Winfrey, M., Elder, J., Krabbenhoft, D., Andren, A., Babiarz, C., Porcella, D.B., and Huckabee, J.W., Sources and fates of mercury and methylmercury in Wisconsin lakes. In: *Mercury as a Global Pollutant: Towards Integration and Synthesis*, C.J. Watras and J.W. Huckabee (Eds.), Lewis Press, Boca Raton, Florida, pp. 153-177 (1994).

WDNR/WDH, *Wisconsin Department of Natural Resources - Wisconsin Division of Health*, Health advisory for people who eat sport fish from Wisconsin water. Madison, Wisconsin (1986).

Weiner, J.G., Fitzgerald, W.F., Watras, C.J., and Rada, R.G., Partitioning and bioavailability of mercury in an experimentally acidified Wisconsin lake. *Environmental Toxicology and Chemistry*, Vol. 9, pp. 909-918 (1990).

List of Clair Patterson's Publications

Brown, H. and Patterson, C.
Composition of meteoritic matter. I: Composition of the silicate phase of stony meteorites. *Journal of Geology* v. **55**, 405-411 (1947).

Brown, H. and Patterson, C.
Composition of meteoritic matter. II: Composition of iron meteorites and of the metal phase of stony meteorites. *Journal of Geology* v. **55**, 508-511 (1947).

Brown, H. and Patterson, C.
The relative abundance of some light nuclear species as determined from the composition of stony meteorites. *Physical Review* v. **72**, 456-457 (1947).

Brown, H. and Patterson, C.
Composition of meteoritic matter. III: Phase equilibria, genetic relationships, and planet structure. *Journal of Geology* v. **56**, 85-111 (1948).

Inghram, M.G., Brown, H., Patterson, C. and Hess, D.C.
The branching ratio of K^{40} radioactive decay. *Physical Review* v. **80**, 916-917 (1950).

Patterson, C.C.
The isotopic composition of trace quantities of lead and calcium: *Atomic Energy Comm., Rept. AECD 3180*, ANL-MG1-33 (1951).

Brown, H., Inghram, M.C., Esper, S.L., Jr., Patterson, C. and Tilton, G.
Isotopic composition of lead and the ages of minerals in a Precambrian granite (abstract). *Bulletin of the Geological Society of America* v. **62**, 1426 (1951).

Tilton, G.R., Patterson, C. and Inghram, M.G.
Mass spectrometric determination of thorium (abstract). *Bulletin of the Geological Society of America* v. **63**, 1305 (1952).

Patterson, C.
The isotopic composition of meteoritic, basaltic and oceanic leads, and the age of the earth. *Report by the Subcommittee on Nuclear Processes in Geological Settings*, National Research Council, National Academy of Sciences, Washington, D.C., 36-40 (1953).

Patterson, C., Brown, H., Tilton, G. and Inghram, M.
Concentration of uranium and lead and the isotopic composition of lead in meteoritic material. *Physical Review* v. **92**, 1234-1235 (1953).

Patterson, C.C., Goldberg, E.D. and Inghram, M.G.
Isotopic compositions of Quaternary leads from the Pacific Ocean. *Bulletin of the Geological Society of America* v. **64**, 1387-1388 (1953).

Patterson, C., Tilton, G. and Inghram, M.
Abundances of uranium and the isotopes of lead in the earth's crust and meteorites (abstract). *Bulletin of the Geological Society of America* v. **64,** 1461 (1953).

Tilton, G.R., Patterson, C.C. and Davis, G.L.
Isotopic composition of lead in olivine bombs (abstract). *Bulletin of the Geological Society of America* v. **65,** 1314-1315 (1954).

Tilton, G.R., Patterson, C., Brown, H., Inghram, M., Hayden, R., Hess, D., and Larsen, E., Jr.
The isotopic composition and distribution of lead, uranium, and thorium in a Precambrian granite. Atomic Energy Commission Report No. AECU-2840, Technical Information Service, Oak Ridge, TN, pp 27 (1954).

Patterson, C.C.
The Pb^{207}/Pb^{206} ages of some stone meteorites. *Geochimica et Cosmochimica Acta* v. **7**, 151-153 (1955).

Patterson, C., Tilton, G. and Inghram, M.
Age of the Earth. *Science* v. **121**, 69-75 (1955).

Aldrich, L.T., Tilton, G.R., Davis, G.L., Nicolaysen, L.O. and Patterson, C.C.
Comparison of U-Pb, Pb-Pb and Pb-Sr ages of Precambrian minerals. *Proceedings of the Geological Association of Canada* v. **7**, Part II, 7-13 (1955).

Tilton, G.R., Patterson, C., Brown, H., Inghram, M., Hayden, R., Hess, D., and Larsen, E., Jr.
Isotopic composition and distribution of lead, uranium, and thorium in a Precambrian granite. *Bulletin of the Geological Society of America* v. **66,** 1131-1148 (1955).

Patterson, C.
Age of meteorites and the earth. *Geochimica et Cosmochimica Acta* v. **10,** 230-237 (1956).

Patterson, C.C.
Age of meteorites (abstract). *Bulletin of the Geological Society of America* v. **67**, 1821 (1956).

Patterson, C., Silver, L. and McKinney, C.
Lead isotopes and magmatic differentiation (abstract). *Proceedings, 20th International Geological Congress*, Section 11-B, 221-222 (1956).

Tilton, G.R. and Patterson, C.C.
The isotopic composition of lead in ultramafic rocks (abstract). *Transactions, American Geophysical Union* v. **37**, 361 (1956).

Chow, T.J. and Patterson, C.C.
Isotopic composition of lead in manganese nodules (abstract). *Bulletin of the Geological Society of America* v. **68**, 1708 (1957).

Engel, A.E.J. and Patterson, C.C.
Isotopic composition of lead in Leadville limestone, hydrothermal dolomite and associated ore. (abstract) *Bulletin of the Geological Society of America* v. **68**, 1723 (1957).

Goldberg, E.D., Patterson, C. and Chow, T.
Ionium-thorium and lead isotope ratios as indicators of oceanic water masses. *Proceedings, United Nations 2nd International Conference on Peaceful Uses of Atomic Energy* v. **18**, 347-350 (1958).

Chow, T.J. and Patterson, C.C.
Lead isotopes in manganese nodules. *Geochimica et Cosmochimica Acta* v. **17**, 21-31 (1959).

Chow, T.J. and Patterson, C.C.
On the primordial lead of the Canyon Diablo meteorite (abstract). *Geokhimiya*, 1124-1125 (1961).

Murthy, V.R. and Patterson, C.C.
Lead isotopes in ores and rocks of Butte, Montana. *Economic Geology*, 59-67 (1961).

Murthy, V.R. and Patterson, C.C.
Reply to discussion by J. Grunig, J. Guilberg and L. Zeihen on origin of Butte ores. *Economic Geology*, v. **56**, 217-218 (1961).

Chow, T.J. and Patterson, C.C.
The occurrence and significance of lead isotopes in pelagic sediments. *Geochimica et Cosmochimica Acta* v. **26**, 263-308 (1962).

Chow, T.J., Tatsumoto, M. and Patterson, C.C.

Lead isotopes and uranium contents in experimental Mohole cores (Guadalupe site). *Journal of Sedimentary Petrology* v. **32**, 866-870 (1962).

Murthy, V.R. and Patterson, C.C.

Primary isochron of zero age for meteorites and the earth. *Journal of Geophysical Research* v. **67**, 1161-1167 (1962) .

Patterson, C.

Characteristics of lead isotope evolution on a continental scale in the earth. *Isotopic and Cosmic Chemistry*, H. Craig, S.L. Miller and G.J. Wasserburg, editors, North-Holland Publishing Co., Amsterdam, 244-268 (1963).

Patterson, C. and Duffield, B.

The isotopic composition of lead in Easter Island rhyolite. *Geochimica et Cosmochimica Acta* v. **27**, 1180-1181 (1963).

Tatsumoto, M. and Patterson, C.C.

The concentration of common lead in sea water. *Earth Science and Meteoritics*, J. Geiss and E.D. Goldberg, editors, North-Holland Publishing Co., Amsterdam, 74-89 (1963).

Tatsumoto, M. and Patterson, C.C.

The concentration of common lead in sea water (Discussed by C.C. Patterson). *Nuclear Geophysics*, Publication 1075, National Academy of Sciences, National Research Council, Washington, D.C., 167-175 (1963).

Tatsumoto, M. and Patterson, C.C.

Concentrations of common lead in some Atlantic and Mediterranean waters and in snow. *Nature* v. **199**, 350-352 (1963).

Patterson C.C.

Preliminary report on the isotopic composition of leads in lavas from the Hawaiian Islands. *Recent Researches in the Fields of Hydrosphere, Atmosphere and Nuclear Geochemistry*, Maruzen Company, Ltd., Tokyo, Japan, 257-261 (1964).

Patterson, C. and Tatsumoto, M.

The significance of lead isotopes in detrital feldspar with respect to chemical differentiation within the earth's mantle. *Geochimica et Cosmochimica Acta* v. **28**, 1-22 (1964).

Ledent, D., Patterson, C. and Tilton, G.R.
Ages of zircon and feldspar concentrates from North American beach and river sands. *Journal of Geology* v. **72**, 112-122 (1964).

Tatsumoto, M. and Patterson, C.
Age studies of zircon and feldspar concentrates from the Franconia sandstone. *Journal of Geology* v. **72**, 232-242 (1964).

Patterson, C.C.
Contaminated and natural lead environments of man. *Archives of Environmental Health* v. **11**, 344-360 (1965).

Patterson, C.C.
Letter to the Editor (Reply). *Archives of Environmental Health* v. **12**, 136-138 (1966).

Patterson, C.C.
Measurement of common lead in Antarctic snow (Field Report). *Antarctic Journal of the United States* v. **1**, 139-140 (1966)

Chow, T.J. and Patterson, C.C.
Concentration profiles of barium and lead in Atlantic waters off Bermuda. *Earth and Planetary Science Letters* v. **1**, 397-400 (1966).

Patterson, C.C. and Salvia, J.D.
Lead in the modern environment. How much is natural? *Scientist and Citizen* v. **10**, 66-79 (1968).

Murozumi, M., Chow, T.J. and Patterson, C.
Chemical concentrations of pollutant lead aerosols, terrestrial dusts and sea salts in Greenland and Antarctic snow strata. *Geochimica et Cosmochimica Acta* v. **33**, 1247-1294 (1969).

Murozumi, M., Chow, T.J. and Patterson, C.
Lead, dust, and salt in firn and ice from Camp Century and Byrd Station. *Antarctic Journal of the United States* v. **4**, 218 (1969).

Patterson, C.C., Chow, T.J. and Murozumi, M.
The possibility of measuring variations in the intensity of worldwide lead smelting during medieval and ancient times using lead aerosol deposits in polar snow strata. *Scientific Methods in Medieval Archaeology*, Rainer Berger, editor, University of California Press, Berekley, 339-350 (1970).

Murozumi, M. and Patterson, C.C.
Neutron activation analysis of chlorine in Arctic and Antarctic snow strata. *Japanese Analyst* v. **8**, 1049-1056 (1970).

Murozumi, M. and Patterson, C.C.
Determination of sodium in Arctic and Antarctic strata. *Japanese Analyst* v. **8**, 1057-1064 (1970).

Patterson, C.C.
Lead in the environment. *Connecticut Medicine* v. **35**, 347-352 (1971).

Patterson, C.
Native copper, silver, and gold accessible to early metallurgists. *American Antiquity* v. **36**, 286-321 (1971).

Patterson, C.
Lead. *Impingement of Man on the Oceans*, D.W. Hood, editor, John Wiley & Sons, Inc., New York, 245-258 (1971).

Patterson, C.
Lead pollution. *Hygiene and Sanitation* (U.S.S.R.) v. **11**, 89-95 (1971).

Dyrssen, D., Patterson, C., Ui, J., and Weichart, G.F.
Inorganic Chemicals. *Report of the Seminar on Methods of Detection, Measurement and Monitoring of Pollutants in the Marine Environment*, FAO Fisheries Reports, No. 99, Supplement v. **1**, 37-52. Fishery Resources Division, Food and Agriculture Organization of the United Nations, Rome (1971).

Patterson, C.C.
Silver stocks and losses in ancient and medieval times. *The Economic History Review*, 2nd Series, v. **XXV**, 205-235 (1972).

Patterson, C.C.
Dwindling stocks of silver, and their relevance to studies of the metal contents of silver coinage. *Methods of Chemical and Metallurgical Investigation of Ancient Coinage*, E.T. Hall and D.M. Metcaff, editors, Royal Numismatic Society Special Publication No. 8, University Press, Oxford, 149-152 (1972).

Patterson, C.
Limiting the contamination of virgin surfaces of firn by engine exhausts from nearby traverses and overflights. *Proceedings of the Colloquium on Conservation Problems in Antarctica*, B.C. Parker, editor, The Allen Press, Inc., Lawrence, Kansas, 43-47 (1972).

Patterson, C.C.
Determination of natural levels of toxic metals in a terrestrial ecosystem. *Nobel Symposium 20, The Changing Chemistry of the Oceans*, Proceedings of the Twentieth Nobel Symposium held 16-20th August 1971 at Aspenagården, Lerum and Chalmers University of Technology, Göteborg, Sweden, D. Dyrssen and D. Jagner, editors, Almqvist and Wiksell, Stockholm, Wiley Interscience Division, John Wiley & Sons, Inc., N.Y., 289-290 (1972).

Patterson, C.C., Weiss, H.V., Koide, M., and Goldberg, E.D.
Reply to Edward M. Dickson's article "Mercury and lead in the Greenland Ice Sheet: a reexamination of the data". *Science* v. **177**, 538 (1972).

Dyrssen, D., Patterson, C., Ui, J., Weichart, G.F.
Inorganic Chemicals. *A Guide to Marine Pollution*, Seminar in conjunction with the FAO Technical Conference on Marine Pollution and its Effects on Resources and Fishing, E.D. Goldberg, editor, Gordon and Breach Science Publishers, New York, 41-58 (1972).

Hinkley, T. and Patterson, C.
Concentrations of metals in a soil moisture film. *Nature Physical Science* v. **246**, 123-124 (1973).

Hirao, Y. and Patterson, C.C.
Lead pollution by gasoline exhausts in a remote canyon in the Sierra Nevada (abstract). *EOS, Transactions, American Geophysical Union* v. **54**, 505 (1973).

Patterson, C.
Lead in seawater. Workshop Meeting Report. *Science* v. **183**, 553-554 (1974).

Patterson, C.C.
Lead. *Encyclopedia of Science and Technology*, 1974 Yearbook, McGraw-Hill Book Co., New York, 257-260 (1974).

Patterson, C. and Settle, D.
Contribution of lead via aerosol deposition to the Southern California Bight. *Journal de Recherches Atmosphériques, Numéro Spécial 8*, "Symp. Int. sur les Exchanges Ocean/Atmosphère de Matière à l'Etat Particulaire", 957-960 (1974).

Patterson, C., Settle, D., Glover, B., Chow, T.J. and Snyder, C.
Determination of the input and transport of pollutant lead in marine environments using isotope tracers. Excerpts published in *Pollutant Transfer to the Marine Environment: Deliberations and Recommendations of the National Science Foundation, International*

Decade of Ocean Exploration Pollutant Transfer Workshop, January 11-12, 1974, R.A. Duce and P.L. Parker, editors (1974).

Chow, T.J., Patterson, C.C. and Settle, D.
Occurrence of lead in tuna. *Nature* v. **251**, 159-161 (1975).

Hirao, Y. and Patterson, C.C.
Lead aerosol pollution in the High Sierra overrides natural mechanisms which exclude lead from a food chain. *Science* v. **184**, 989-992 (1975).

Brewer, P., Frew, N., Cutshall, N., Wagner, J.J., Duce, R.A., Walsh, P.R, Hoffmann, G.L., Dutton, J.W.R., Fitzgerald, W., Hunt, C.D., Girvin, D.C., Clem, R.G., Patterson, C., Settle, D., Glover, B., Presley, B.J., Trefry, J., Windom, H., Smith, R., Chow, T.J., Doe, B., Giam, C., Martin, J., Parker, P., Perkins, R. and Robertson, D. (Lead in Seawater Workshop participants and advisors).
Interlaboratory lead analyses of standardized samples of seawater. *Marine Chemistry* v. **2**, 69-84 (1975).

Elias, R., Hirao, Y. and Patterson, C.C.
Impact of Present Levels of Aerosol Pb Concentrations on both Natural Ecosystems and Humans. *International Conference on Heavy Metals in the Environment*, Toronto, Ontario, Canada, 257-272 (1975)

Patterson, C.C. and Settle, D.M.
The reduction of order of magnitude errors in lead analyses of biological materials and natural waters by evaluating and controlling the extent and sources of industrial lead contamination introduced during sample collecting, handling, and analysis. *Accuracy in Trace Analysis: Sampling, Sample Handling, and Analysis*, P. LaFleur, editor, National Bureau of Standards Special Publication v. **422**, 321-351 (1976).

Patterson, C.C. and Settle, D.
Interlaboratory analyses of references. *Strategies for Marine Pollution Monitoring*, E.D. Goldberg, editor, John Wiley and Sons, Inc., New York, 235-239 (1976).

Patterson, C., Settle, D. and Glover, B.
Analysis of lead in polluted coastal seawater. *Marine Chemistry* v. **4**, 305-319 (1976).

Patterson, C., Settle, D., Schaule, B., and Burnett, M.
Transport of pollutant lead to the oceans and within ocean ecosystems. *Marine Pollutant Transfer*, H.L. Windom and R.A. Duce, editors, Lexington Books, D.C. Heath and Co., Lexington, MA, 23-38 (1976).

Patterson, C.C., Settle, D.M. Schaule, B.K., and Burnett, M.W.

Standardization of reference samples for certain trace metals: the first step in a multistage program designed to insure the reliability and significance of baseline data for trace metals. *Baseline Studies of Outer Continental Shelf, Southern California* 3 *Report 4.4*, Bureau of Land Management Contract #696-16, U.S. Dept. of Interior, Washington, D.C. (1976).

Burnett, M.W., Settle, D.M. and Patterson, C.C.

Comparative distributions of alkaline earths and lead among tissues of marine plants and animals. *7th International Mass Spectrometry Conference, Florence, Italy, 1976, Extended Abstracts, Paper 152-L* (1976).

Berand, P., Bewers, J.M., Duce, R., Fitzgerald, W.F., Hartling, S., Hunt, C. D. Patterson, C., Settle, D., Smith, R., Walsh, P., Windom, H. and Wong, C.S.

Comparison determinations of lead by investigators analyzing individual samples of seawater in both their home laboratory and in an isotope dilution standardization laboratory. *Marine Chemistry* v. **4**, 389-392 (1976).

Elias, R.W., Hinkley, T.K., Hirao, Y. and Patterson, C.C.

Improved techniques for studies of mass balances and fractionations among families of metals within terrestrial ecosystems. *Geochimica et Cosmochimica Acta* v. **40**, 583-587 (1976).

Patterson, C.C.

Age of the earth. *Encyclopedia of Science and Technology* 5, McGraw-Hill Book Co., New York, 367-369 (1977).

Patterson, C. and Settle, D.

Comparative distributions of alkalies, alkaline earths and lead among major tissues of the tuna *Thunnus alalunga*. *Marine Biology v.* **39**, 289-295 (1977).

Elias, R., Hirao, Y. and Patterson, C.

Impact of present levels of aerosol Pb concentration on both natural ecosystems and humans. *International Conference on Heavy Metals in the Environment, Toronto, 1975* v. **2**, Part 1, T.C. Hutchison, editor, 257-272 (1977).

Patterson, C.C.

Review of *LEAD. Environmental Health Criteria 3. The Quarterly Review of Biology* v. **53**, 483-484 (1978).

Patterson, C. and Settle, D.

Comparison of isotope dilution mass spectrometric techniques with relative analytical methods for the analysis of trace metals in biological materials. Presented at American Society for Mass Spectrometry Symposium, St. Louis, MO (1978).

Burnett, M.W., Settle, D.M. and Patterson, C.C.

Comparative distributions of alkaline earths and Pb among tissues of marine plants and animals. *Advances in Mass Spectrometry* v. **7B**, N.R. Daly, editor, Heyden and Son, Ltd., for Institute of Petrology, London, 1697-1701 (1978).

Elias, R., Hirao, Y. and Patterson, C.C.

The deposition of lead aerosols on plant surfaces in a remote subalpine ecosystem. *Environmental Chemistry and Cycling Processes*, D.C. Adriano and I.L. Brisbin, Jr., editors, DOE pub. no. CONF-760429, 691-699 (1978).

Fitzgerald, W., Settle, D, Unni, C., Patterson, C. and Duce, R.

Lead and mercury in remote South Pacific rain (abstract). *EOS, Transactions, American Geophysical Union* v. **59**, 1078 (1978).

Unni, C., Fitzgerald, W., Settle, D., Gill, G., Ray, B., Patterson, C. and Duce, R.

The impact of volcanic emissions on the global atmospheric cycles of sulfur, mercury and lead (abstract). *EOS, Transactions, American Geophysical Union* v. **59**, 1223 (1978).

Patterson, C.

Discussion of "Sulfate in Antarctic Snow: Spatio-Temporal Distribution", by R. Delmas and C. Boutron. *Atmospheric Environment* v. **13**, 1063 (1979).

Ericson, J.E., Shirahata H., and Patterson, C.C.

Skeletal concentrations of lead in ancient Peruvians. *The New England Journal of Medicine* v. **300**, 946-951 (1979).

Patterson, C.C.

An alternative perspective -- Lead pollution in the human environment: origin, extent and significance. *Lead in the Human Environment, A report prepared by the Committee on Lead in the Human Environment*, National Research Council, National Academy of Sciences, National Academy Press, Washington, D.C., 265-349 (1980).

Burnett, M. and Patterson, C.

Analysis of natural and industrial lead in marine ecosystems. *Lead in the Marine Environment, Proc. Int. Experts Discussion*, Rovinj 1977, M. Branika and Z. Konrad, editors, Pergamon Press, Oxford, England, 7-13 (1980).

Burnett, M.W. and Patterson, C.C.

Perturbation of natural lead transport in nutrient calcium pathways of marine ecosystems by industrial lead. *Isotope Marine Chemistry*, E.D. Goldberg, Y. Horibe and K Saruhashi, editors, Uchida Rokakuho Publishing Co. Ltd., 413-438 (1980).

Burnett, M., Ng, A., Settle, D. and Patterson, C.

Impact of man on coastal marine ecosystems. *Lead in the Marine Environment, Proc. Int. Experts Discussion*, Rovinj 1977, M. Branika and Z. Konrad, editors, Pergamon Press, Oxford, England, 15-30 (1980).

Elias, R.W. and Patterson, C.C.

The toxicological implications of biogeochemical studies of atmospheric lead. *Polluted Rain*, T.Y. Toribara, M.W. Miller and P.C. Morrow, editors, Plenum Publishing Corp., New York, 391-403 (1980).

Everson, J. and Patterson, C.C.

"Ultra-clean" isotope dilution/mass spectrometric analyses for lead in human blood plasma indicate that most reported values are artificially high. *Clinical Chemistry* v. **26**, 1603-1607 (1980).

Ng, A.C. and Patterson, C.C.

Chronological variations in barium concentration, lead concentration and lead isotopic composition in sediment of four Southern California off-shore basins. *Southern California Baseline Study Benthic*, Year Two II *Section 14*, III *Section 14*, IV *Section 14*, Bureau of Land Management, Department of Interior Contract AA550-Ct6-40 (1980).

Schaule, B.K. and Patterson, C.

The occurrence of lead in the Northeast Pacific, and the effects of anthropogenic inputs. *Lead in the Marine Environment, Proc. Int. Experts Discussion*, Rovinj 1977, M. Branika and Z. Konrad, editors, Pergamon Press, Oxford, England, 31-43 (1980).

Settle, D.M. and Patterson, C.C.

Lead in albacore: guide to lead pollution in Americans. *Science* v. **207**, 1167-1176 (1980).

Shirahata, H., Elias, R.W., Patterson, C.C. and Koide, M.

Chronological variations in concentrations and isotopic compositions of anthropogenic atmospheric lead in sediments of a remote subalpine pond. *Geochimica et Cosmochimica Acta* v. **44**, 149-162 (1980).

Patterson, C.C.

Acceptance speech for the V.M. Goldschmidt Medal. *Geochimica et Cosmochimica Acta* v. **45**, 1383-1388 (1981).

Patterson, C.C.

Tracking down a commonplace poison. *The Grinnell Magazine XIII*, v. **5**, 18-19 (1981).

Everson, J. and Patterson, C.C.

Plasma lead values: are they accurate? Reply to letter to the editor. *Clinical Chemistry* v. **27**, 765-766 (1981).

Flegal, A.R., Schaule, B.K., Settle, D.M. and Patterson, C.C.

Lead concentrations and stable isotopic compositions in surface waters of the Central Pacific (abstract). *EOS, Transactions, American Geophysical Union v.* **62**, 894 (1981).

Ng, A., and Patterson, C.C.

Natural concentrations of lead in ancient Arctic and Antarctic ice. *Geochimica et Cosmochimica Acta* v. **45**, 2109-2121 (1981).

Schaule, B.K. and Patterson, C.C.

Lead concentrations in the northeast Pacific: evidence for global anthropogenic perturbations. *Earth and Planetary Science Letters* v. **54**, 97-116 (1981).

Elias, R.W., Hirao, Y. and Patterson, C.C.

The circumvention of natural biopurification of calcium along nutrient pathways by atmospheric inputs of industrial lead. *Geochimica et Cosmochimica Acta* v. **46**, 2561-2580 (1982).

Patterson, C.C., Boutron, C. and Flegal, R.

Present status and future of lead studies in polar snow. *Greenland Ice Core: Geophysics, Geochemistry, and the Environment* (1982 Proceedings of NSF Greenland Ice Sheet Program), Geophysical Monograph #33, C.C. Langway, Jr., H. Oeschger and W. Dansgaard, editors, American Geophysical Union, Washington, D.C., 101-104 (1982).

Ericson, J.E., Pandolfi, L. and Patterson, C.C.

Pyrotechnology of copper extraction: methods of detection and implications. *Early Pyrotechnology* The Evolution of the First Fire-Using Industries, T.A. Wertime and S. Wertime, editors, Smithsonian Institution Press, GPO, Washington, D.C. pp. 192-203 (1982).

Ng, A. and Patterson, C.C.
Changes of lead and barium with time in California off-shore basin sediments. *Geochimica et Cosmochimica Acta* v. **46**, 2307-2321 (1982).

Settle, D.M. and Patterson, C.C.
Magnitudes and sources of precipitation and dry deposition fluxes of industrial and natural leads to the North Pacific at Enewetak. *Journal of Geophysical Research* v. **87**, 8857-8869 (1982).

Settle, D.M. and Patterson, C.C.
Magnitudes of eolian lead inputs to the oceans control variations in vertical lead concentration profiles in the oceans. *SEAREX Newsletter* v. **5** (2), 21-26, Published by the Graduate School of Oceanography, University of Rhode Island, Narragansett, RI 02882 (1982).

Settle, D.M., Patterson, C.C., Turekian, K.K. and Cochran, J.K.
Lead precipitation fluxes at tropical oceanic sites determined from ^{210}Pb measurements. *Journal of Geophysical Research* v. **87**, 1239-1245, correction: 3169-3171 (1982).

Patterson, C.C.
Criticism of "Flow of metals into the global atmosphere" [Jaworowski et al., *Geochimica et Cosmochimica Acta* v. **45**, 2185-2199 (1981)] *Geochimica et Cosmochimica Acta* v. **47**, 1163-1168 (1983).

Patterson, C.C.
British Mega Exposures to Industrial Lead. *Lead Versus Health* (International Symposium on Low Level lead Exposure and Its Effects on Human Beings, London, 1982), M. Rutter and R. R. Jones, editors, John Wiley & Sons, Ltd, Chichester, 17-32 (1983).

Patterson, C.C.
Natural levels of lead in humans. *Carolina Environmental Essay Series III,* The Institute for Environmental Studies, The University of North Carolina, Chapel Hill (1983).

Boutron, C.F. and Patterson, C.C.
The occurrence of lead in Antarctic recent snow, firn deposited over the last two centuries and prehistoric ice. *Geochimica et Cosmochimica Acta* v. **47**, 1355-1368 (1983).

Flegal, A.R. and Patterson, C.C.
Vertical concentration profiles of lead in the Central Pacific at 15 N and 20 S. *Earth and Planetary Sciences Letters* v. **64**, 19-32 (1983).

Schaule, B.K. and Patterson, C.C.

Perturbations of the natural lead depth profile in the Sargasso Sea by industrial lead. *Trace Metals in Sea Water, NATO Conference Series IV: Marine Sciences,,* C.S. Wong, E. Boyle, K.W. Bruland, J.D. Burton and E.D. Goldberg, editors, Plenum Press, New York, v. **9**, 487-503 (1983).

Flegal, A.R., Itoh, K. and Patterson, C.C.

Concentrations and isotopic compositions of lead in biologically productive coastal waters of the Northeast Pacific (abstract). *EOS, Transactions, American Geophysical Union* v. **64**, 734-735 (1983).

Flegal, A.R. and Patterson, C.C.

Lead in plankton: Bioenrichment at the primary and secondary trophic levels (abstract). *EOS, Transactions, American Geophysical Union* v. **65**, 924 (1984).

Flegal, A.R., Schaule, B.K. and Patterson, C.C.

Stable isotopic ratios of lead in surface waters of the Central Pacific. *Marine Chemistry* v. **14**, 281-287 (1984).

Patterson, C.C. and Settle, D.M.

Input of sea water lead to marine air and rain by sea spray (abstract). *EOS, Transactions, American Geophysical Union,* v. **66**, 1318 (1985).

Patterson, C., Settle, D. and Rosman, K.

Recycled lead in sea spray salt. *SEAREX Newsletter* 8 (1), 8-10, Published by the Graduate School of Oceanography, University of Rhode Island, Narragansett, RI 02882 (1985).

Flegal, A.R. and Patterson, C.C.

Thallium concentrations in seawater. *Marine Chemistry* v. **15**, 327-331 (1985).

Settle, D. and Patterson C.

New Zealand results. *SEAREX Newsletter* v. **8**, (1), 10-11, Published by the Graduate School of Oceanography, University of Rhode Island, Narragansett, RI 02882 (1985).

Flegal, A.R., Martin, J.H. and Patterson, C.C.

Lead fluxes in the Northeast Pacific (abstract). *EOS, Transactions, American Geophysical Union* v. **66**, 1290 (1985).

Patterson, C.C.
Harrison Brown's influence on twentieth-century developments in geochronology. *Earth and the Human Future: Essays in Honor of Harrison Brown*, Kirk R. Smith, Fereidun Fesharaki and John Holdren, editors, Westview Press, Boulder, Colorado (1986).

Patterson, C.C. and Settle, D.M.
Magnitude of lead flux to the atmosphere from volcanoes. *Proceedings of the International Volcanological Congress*, Auckland, New Zealand (1986).

Boutron, C. and Patterson C.C.
Changes of lead concentrations in Antarctic ice during the transition from Late Wisconsin to Holocene. *Nature* v. **323**, 222-225 (1986).

Flegal, A.R., Settle, D.M. and Patterson, C.C.
Thallium in marine plankton. *Marine Biology* v. **90**, 501-503 (1986).

Flegal, A.R., Itoh, K., Patterson, C.C. and Wong, C.S.
Vertical profile of lead isotopic compositions in the Northeast Pacific. *Nature* v. **321**, 689-690 (1986).

Patterson, C.C.
News and Views: Global pollution measured by lead in mid-ocean sediments. *Nature* v. **326**, 244-245 (1987).

Patterson, C.C. and Settle, D.M.
Eolian fluxes of industrial and natural lead to the seas at various geographic locations. *SEAREX Newsletter* v. **10** (1), 19-26, Published by the Graduate School of Oceanography, University of Rhode Island, Narragansett, RI 02882 (1987).

Patterson, C.C. and Settle, D.M.
Magnitude of lead flux to the atmosphere from volcanoes. *Geochimica et Cosmochimica Acta* v. **51**, 675-681 (1987).

Patterson, C.C. and Settle, D.M.
Review of data on eolian fluxes of industrial and natural lead to the lands and seas in remote regions on global scale. *Marine Chemistry* v. **22**, 137-162 (1987).

Patterson, C.C., Shirahata, H. and Ericson, J.E.
Lead in ancient human bones and its relevance to historical developments of social problems with lead. *Science of the Total Environment* v. **61**, 167-200 (1987).

Boutron, C.F. and Patterson, C.C.

Relative levels of natural and anthropogenic lead in recent Antarctic snow. *Journal of Geophysical Research* v. **92**, 8454-8464 (1987).

Boutron, C.F., Patterson, C.C., Petrov, V.N. and Barkov, N.I.

Preliminary data on changes of lead concentrations in Antarctic ice from 155,000 to 26,000 years BP *Atmospheric Environment* v. **21**, 1197-1202 (1987).

Maring, H., Settle, D.M., Buat-Ménard, P., Dulac, F. and Patterson, C.C.

Stable lead isotope tracers of air mass trajectories in the Mediterranean region. Nature v. **330**, 154-156 (1987).

Boutron, C.F., Patterson, C.C., Lorius, C., Petrov, V.N. and Barkov, N.I.

Atmospheric lead in Antarctic ice during the last climatic cycle. *Annals of Glaciology* v. **10**, 5-9 (1988).

Patterson, C.C., Boutron, C.F. and Petrov, V.N.

Lead records in Antarctic ice: Changes in global atmospheric concentrations during the past 150,000 years. *The Antarctic Journal 1987 Review Issue* v. **22**, 85-87 (1988).

Boutron, C.F., Patterson, C.C. and Barkov, N.I.

Assessing the quality of thermally drilled deep Antarctic ice cores for trace elements analysis. *Ice Core Drilling* (C. Rado and D. Beaudoing, editors), Proceedings of the Third International Workshop on Ice Drilling Technology (1988).

Maring, H., Patterson, C.C. and Settle, D.M.

Atmospheric input fluxes of industrial and natural Pb from the westerlies to the Mid-North Pacific. *Chemical Oceanography* v. **10**, 83-106 (special SEAREX volume) (1989).

Rosman, K.J.R., Patterson, C.C. and Settle, D.M.

The distribution of lead between sea salt, dust, and lead-rich aerosols collected on a cascade impactor in the mid South Pacific easterlies air at American Samoa. *Journal of Geophysical Research* v. **95**, 3687-3691 (1990).

Erel, Y., Patterson, C.C., Scott, M.J. and Morgan, J.J.

Transport of industrial lead in snow through soil to stream water and groundwater. *Chemical Geology* v. **85**, 383-392 (1990).

Bolshov, M.A., Koloshnikov, V, Boutron, C., Patterson, C.C., and Barkov, N.
Determination of toxic metals in Environmental Objects by Laser Excited Atomic Fluorescence Spectrometry. *Optoelectronics for Environmental Science,* Plenum Press, NY (1990).

Ericson, J.E., Han, D. and Patterson, C.C.
Biogeochemical analysis of fossil teeth from Giganto Pithecus Cave, Luicheng, Kwangsi Province, China. *Vertebrata Palasiatica.* (1990)

Boutron, C.F., Bolshov, M.A., Koloshnikov, V.G., Patterson, C.C. and Barkov, N.I.
Direct determination of lead in Vostok Antarctic ancient ice by laser excited atomic fluorescence spectrometry. *Atmospheric Environment* v. **24A**, 1797-1800 (1990).

Boutron, C.F., Patterson, C.C. and Barkov, N.I.
The occurrence of zinc in Antarctic ancient ice and recent snow. *Earth and Planetary Science Letters* v. **101**, 248-259 (1990).

Erel, Y., Patterson, C.C., Scott, M.J., and Morgan, J.J.
Transport of Industrial Lead in Snow Through Soil to Stream Water and Groundwater. *Chemical Geology* v. **85**, 383-392 (1990).

Boutron, C.F., Patterson, C.C., and Barkov, N.I.
The Occurrence of Zinc in Antarctic Ancient Ice and Recent Snow. *Earth and Planetary Science Letters,* v. **101**, 248-259 (1990).

Patterson, C.C.
Forward to book entitled *Rome's Ruin by Lead Poison* by S. Colum Gilfillan, *Wenzel Press* (1990).

Patterson, C., Ericson, J., Manea-Krichten, M. and Shirahata, H.
Natural skeletal levels of lead in *Homo sapiens sapiens* uncontaminated by technological lead. *Science of the Total Environment* v. **107**, 205-236 (1991).

Manea-Krichten, M., Patterson, C., Miller, G., Settle, D. and Erel, Y.
Comparative increases of lead and barium with age in human tooth enamel, rib and ulna. *Science of the Total Environment* v. **107**, 179-203 (1991).

Church, T.M., Véron, A., Patterson, C.C., Settle, D., Erel, Y., Maring, H.R. and Flegal, A.R.
Trace elements in the North Atlantic Troposphere: shipboard results of precipitation and aerosols. *Global Biogeochemical Cycles 4,* v. **4**, 431-443 (1991).

Settle, D.M. and Patterson, C.C.

Eolian inputs of lead to the South Pacific via rain and dry deposition from industrial and natural sources. *Stable Isotope Geochemistry*: A Tribute to Samuel Epstein (H.P. Taylor, Jr., J.R. O'Neil and I.R. Kaplan, editors), Geochemical Society Special Publication No. 3. 285-294 (1991).

Erel, Y. Morgan, J.J., and Patterson, C.C.

Natural Levels of Lead and Cadmium in a Remote Mountain Stream. *Geochimica et Cosmochimica Acta*, v. **55**, 707-719 (1991).

Church, T., Véron, A., Patterson, C.C. and Settle, D.

Trace Metal Scavenging from the North Atlantic Troposhere, *Precipitation Scavenging and Atmosphere-Surface Exchange*, v. **1**, The Georgii Volume, Precipitation Scavenging Process, coordinated by S.E. Schwartz and W.G.N. Slinn, *Proceedings of the Fifth International Conference on Precipitation Scavenging and Atmosphere-Surface Exchange*, Richland, Washington, 443-446 (1991)

Veron, A., Church, T.M., Patterson, C.C., Erel, Y., and Merrill, J.T.

Continental Origin and Industrial Sources of Trace-Metals in the Northwest Atlantic Troposphere. *Journal of Atmospheric Chemistry*, v. **14**, 339-351 (1992).

Bolshov, M.A., Koloshnikow, V.G., Rudnev, S.N., Boutron, C.F., Görlach, U., and Patterson, C.C.

Detection of Trace Amounts of Toxic Metals in Environmental Samples by Laser-excited Atomic Fluorescence Spectrometry. *Journal of Analytical Atomic Spectrometry*, v. **7**, 99-104 (1992).

Boutron C.F., Rudniev S.N., Bolshov M.A., Koloshnikov V.G., Patterson, C.C., and Barkov, N.I.

Changes in Cadmium Concentrations in Antarctic Ice and Snow during the Past 155,000 Years. *Earth and Planetary Science Letters*, v. **117**, 431-441 (1993).

Patterson, C.C., and Settle, D.

New Mechanisms in Lead Biodynamics at Ultra-low Levels. *NeuroToxicology*, v.**14**, 291-300 (1993).

Véron, A.J., Church, T.M., Flegal, A.R., Patterson, C.C. and Erel, Y.

Response of Lead Cycling in the Surface Sargasso Sea to Changes in Tropospheric Input. *J. Geophys. Res.*, v. **98**, 18,269-18,276 (1993).

Patterson, C.C.

Historical changes in integrity and worth of scientific knowledge. *Geochimica et Cosmochimica Acta*, v. **58,** n. 15, pp. 3141-3143 (1994).

Patterson, C.C.

Delineation of Separate Brain Regions Used for Scientific vs. Engineering Modes of Thinking, *Geochimica et Cosmochimica Acta*, v. **58**, 3321-3327 (1994).

Vérnon, A.J., Church, T.M., Patterson, C.C., and Flegal, A.R.

Use of stable lead isotopes to characterize the sources of anthropogenic lead in North America surface waters. *Geochimica et Cosmochimica Acta*, v. **58,** 3191-3198 (1994).

Hong, S, Candelone, JP, Patterson, CC, & Boutron, C.F.

Greenland Ice Evidence of Hemispheric Pollution for Lead Two Millennia Ago by Greek & Roman Civilizations. *Science,* v. **265,** 1841-1843 (1994).

Rosman, K.J.R., Chisholm, W. Boutron, C.F., Candelone, J.P. and Patterson, C.C.

Isotopic Evidence for Anthropogenic Lead in Antarctic Snow and Ice. *Geophysical Research Letters,* **v. 21**, pp. 2269-2672 (1994).

Murozumi, M., Nakamura, S., and Patterson, C.C.

Industrial Lead Contamination of Natural Ecosystems in Japan, *American Geophysical Union,* Geophysical Monograph Series volume on "Isotopic Studies of Crust-Mantle Evolution" (1995).

Index

U

V

W

Z